GEOMETRY DEMYSTIFIED

STAN GIBILISCO

McGRAW-HILL
New York Chicago San Francisco Lisbon London
Madrid Mexico City Milan New Delhi San Juan
Seoul Singapore Sydney Toronto

The McGraw-Hill Companies

Cataloging-in-Publication Data is on file with the Library of Congress

To Samuel, Tim, and Tony
from Uncle Stan

1 2 3 4 5 6 7 8 9 0 DOC/DOC 0 9 8 7 6 5 4 3

ISBN 0-07-141650-1

The sponsoring editor for this book was Scott L. Grillo and the production supervisor was Sherri Souffrance. It was set in Times Roman by Keyword Publishing Services Ltd.

Printed and bound by RR Donnelley.

This book was printed on recycled, acid-free paper containing a minimum of 50% recycled, de-inked fiber.

CONTENTS

CONTENTS

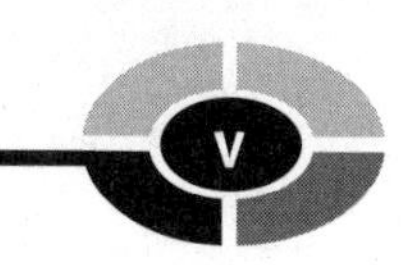

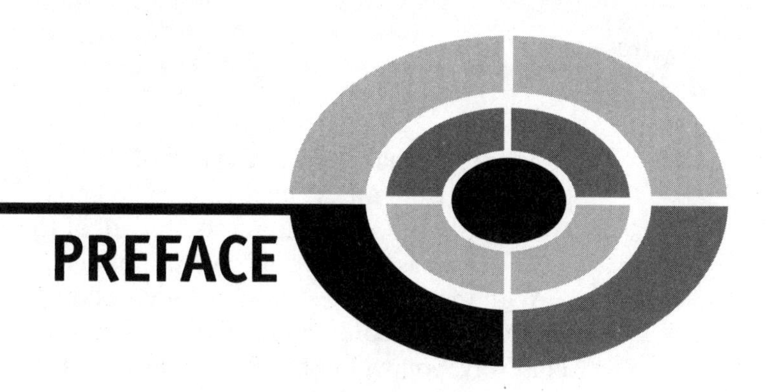

PREFACE

This book is for people who want to get acquainted with the concepts of basic geometry without taking a formal course. It can serve as a supplemental text in a classroom, tutored, or home-schooling environment. It should also be useful for career changers who need to refresh their knowledge of the subject. I recommend that you start at the beginning of this book and go straight through.

This is not a rigorous course in theoretical geometry. Such a course defines *postulates* (or *axioms*) and provides deductive proofs of statements called *theorems* by applying mathematical logic. Proofs are generally omitted in this book for the sake of simplicity and clarity. Emphasis here is on practical aspects. You should have knowledge of middle-school algebra before you begin this book.

This introductory work contains an abundance of practice quiz, test, and exam questions. They are all multiple-choice, and are similar to the sorts of questions used in standardized tests. There is a short quiz at the end of every chapter. The quizzes are "open-book." You may (and should) refer to the chapter texts when taking them. When you think you're ready, take the quiz, write down your answers, and then give your list of answers to a friend. Have the friend tell you your score, but not which questions you got wrong. The answers are listed in the back of the book. Stick with a chapter until you get most of the answers correct.

This book is divided into two sections. At the end of each section is a multiple-choice test. Take these tests when you're done with the respective sections and have taken all the chapter quizzes. The section tests are "closed-book," but the questions are not as difficult as those in the quizzes. A satisfactory score is three-quarters of the answers correct. Again, answers are in the back of the book.

There is a final exam at the end of this course. It contains questions drawn uniformly from all the chapters in the book. Take it when you have finished both sections, both section tests, and all of the chapter quizzes. A satisfactory score is at least 75 percent correct answers.

With the section tests and the final exam, as with the quizzes, have a friend tell you your score without letting you know which questions you missed. That way, you will not subconsciously memorize the answers. You can check to see where your knowledge is strong and where it is not.

I recommend that you complete one chapter a week. An hour or two daily ought to be enough time for this. When you're done with the course, you can use this book, with its comprehensive index, as a permanent reference.

Suggestions for future editions are welcome.

Acknowledgments

Illustrations in this book were generated with *CorelDRAW*. Some clip art is courtesy of Corel Corporation, 1600 Carling Avenue, Ottawa, Ontario, Canada K1Z 8R7.

I extend thanks to Emma Previato of Boston University, who helped with the technical editing of the manuscript for this book.

STAN GIBILISCO

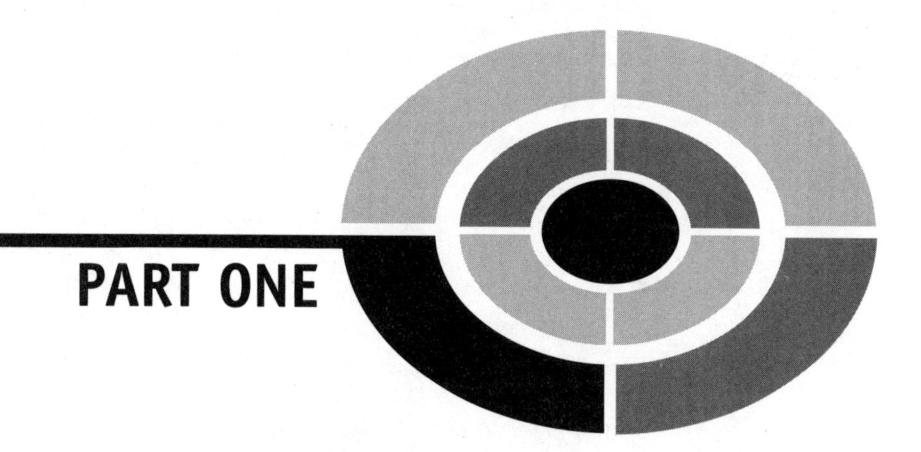

PART ONE

Two Dimensions

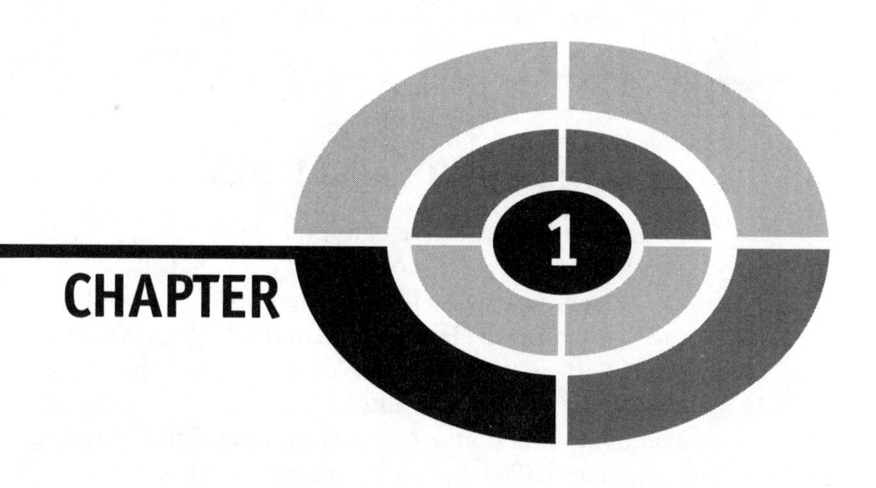

Some Basic Rules

The fundamental rules of geometry go all the way back to the time of the ancient Egyptians and Greeks, who used geometry to calculate the diameter of the earth and the distance to the moon. They employed the laws of *Euclidean geometry* (named after Euclid, a Greek mathematician who lived in the 3rd century B.C.). Euclidean *plane geometry* involves points and lines on perfectly flat surfaces.

Points and Lines

In plane geometry, certain starting concepts aren't defined formally, but are considered intuitively obvious. The *point* and the *line* are examples. A point can be envisioned as an infinitely tiny sphere, having height, width, and depth all equal to zero, but nevertheless possessing a specific location. A line can be thought of as an infinitely thin, perfectly straight, infinitely long wire.

NAMING POINTS AND LINES

Points and lines are usually named using uppercase, italicized letters of the alphabet. The most common name for a point is P (for "point"), and the most common name for a line is L (for "line"). If multiple points are involved in a scenario, the letters immediately following P are used, for example Q, R, and S. If two or more lines exist in a scenario, the letters immediately following L are used, for example M and N. Alternatively, numeric subscripts can be used with P and L. Then we have points called P_1, P_2, P_3, and so forth, and lines called L_1, L_2, L_3, and so forth.

TWO POINT PRINCIPLE

Suppose that P and Q are two different geometric points. Two distinct points define one and only one (that is, a unique) line L. The following two statements are always true, as shown in Fig. 1-1:

- P and Q lie on a common line L
- L is the only line on which both points lie

Fig. 1-1. Two point principle. For two specific points P and Q, line L is unique.

DISTANCE NOTATION

The distance between any two points P and Q, as measured from P towards Q along the straight line connecting them, is symbolized by writing PQ. Units of measurement such as meters, feet, millimeters, inches, miles, or kilometers are not important in pure mathematics, but they are important in physics and engineering. Sometimes a lowercase letter, such as d, is used to represent the distance between two points.

LINE SEGMENTS

The portion of a line between two different points P and Q is called a *line segment*. The points P and Q are called the *end points*. A line segment can theoretically include both of the end points, only one of them, or neither of them.

If a line segment contains both end points, it is a *closed line segment*. If it contains one of the end points but not the other, it is a *half-open line segment*. If it contains neither end point, it is an *open line segment*. Whether a line segment is closed, half-open, or open, its length is the same. Adding or taking away a single point makes no difference, mathematically, in the length because points have zero size in all dimensions! Yet the conceptual difference between these three types of line segments is like the difference between daylight, twilight, and darkness.

RAYS (HALF LINES)

Sometimes, mathematicians talk about the portion of a geometric line that lies "on one side" of a certain point. In Fig. 1-1, imagine the set of points that starts at P, then passes through Q, and extends onward past Q forever. This is known as a *ray* or *half line*.

The ray defined by P and Q might include the end point P, in which case it is a *closed-ended ray*. If the end point is left out, the theoretical object is an *open-ended ray*. In either case, the ray is said to "begin" at point P; informally we might say that it is either "tacked down at the end" or "dangling at the end."

MIDPOINT PRINCIPLE

Suppose there is a line segment connecting two points P and R. Then there is one and only one point Q on the line segment such that $PQ = QR$, as shown in Fig. 1-2.

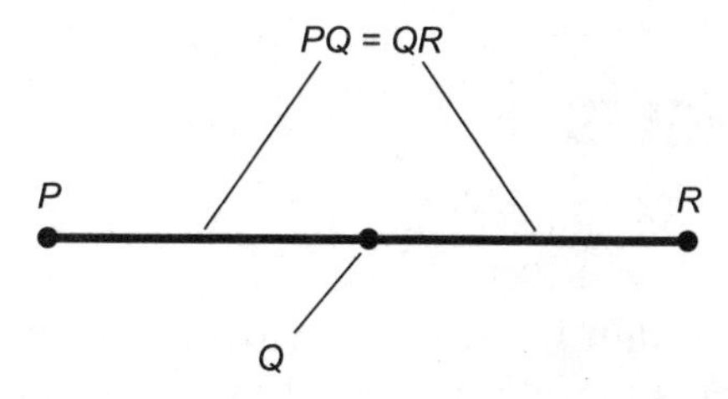

Fig. 1-2. Midpoint principle. Point Q is unique.

PROBLEM 1-1

Suppose, in Fig. 1-2, we find the midpoint Q_2 between P and Q, then the midpoint Q_3 between P and Q_2, then the midpoint Q_4 between P and Q_3, and so on. In mathematical language, we say we keep finding midpoints $Q_{(n+1)}$

between P and Q_n, where n is a positive whole number. How long can this process go on?

SOLUTION 1-1

The process can continue forever. In theoretical geometry, there is no limit to the number of times a line segment can be cut in half. This is because a line segment contains an infinite number of points.

PROBLEM 1-2

Suppose we have a line segment with end points P and Q. What is the difference between the distance PQ and the distance QP?

SOLUTION 1-2

This is an interesting question. If we consider distance without paying attention to the direction in which it is measured, then $PQ = QP$. But if direction is important, we define $PQ = -QP$.

In basic plane geometry, direction is sometimes specified in diagrams in order to get viewers to move their eyes from right to left instead of from left to right, or from bottom to top rather than from top to bottom.

Angles and Distances

When two lines intersect, four *angles* exist at the point of intersection. Unless the two lines are perpendicular, two of the angles are "sharp" and two are "dull." When the two lines are perpendicular, each of the four angles is a *right angle*. Angles can also be defined by sets of three points when the points are connected by line segments.

MEASURING ANGLES

The two most common units of angular measure are the *degree* and the *radian*.

The degree (°) is the unit familiar to lay people. One degree (1°) is 1/360 of a full circle. This means that 90° represents a quarter circle, 180° represents a half circle, 270° represents three-quarters of a circle, and 360° represents a full circle.

A *right angle* has a measure of 90°, an *acute angle* has a measure of more than 0° but less than 90°, and an *obtuse angle* has a measure of more than 90° but less than 180°. A *straight angle* has a measure of 180°. A *reflex angle* has a measure of more than 180° but less than 360°.

The radian (rad) is defined as follows. Imagine two rays emanating outward from the center point of a circle. Each of the two rays intersects the circle at a point; call these points P and Q. Suppose the distance between P and Q, as measured along the arc of the circle, is equal to the radius of the circle. Then the measure of the angle between the rays is one radian (1 rad). There are 2π radians in a full circle, where π (the lowercase Greek letter pi, pronounced "pie") stands for the ratio of a circle's circumference to its diameter. The value of π is approximately 3.14159265359, often rounded off to 3.14159 or 3.14.

A right angle has a measure of $\pi/2$ rad, an acute angle has a measure of more than 0 rad but less than $\pi/2$ rad, and an obtuse angle has a measure of more than $\pi/2$ rad but less than π rad. A straight angle has a measure of π rad, and a reflex angle has a measure larger than π rad but less than 2π rad.

ANGLE NOTATION

Imagine that P, Q, and R are three distinct points. Let L be the line segment connecting P and Q; let M be the line segment connecting R and Q. Then the angle between L and M, as measured at point Q in the plane defined by the three points, can be written as $\angle PQR$ or as $\angle RQP$, as shown in Fig. 1-3.

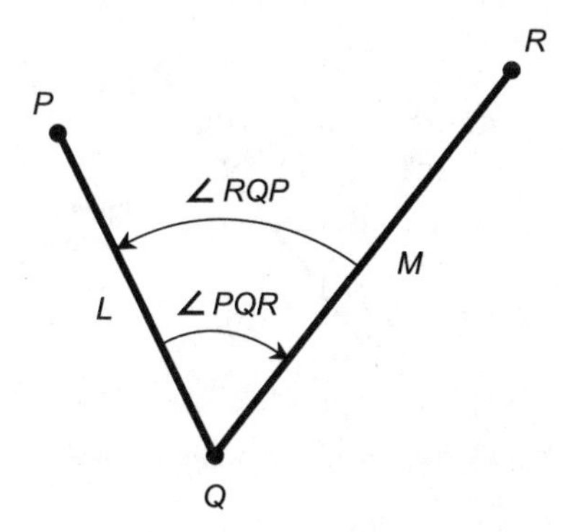

Fig. 1-3. Angle notation.

If the rotational sense of measurement is specified, then $\angle PQR$ indicates the angle as measured from L to M, and $\angle RQP$ indicates the angle as measured from M to L. If rotational sense is important, counterclockwise is usually considered positive, and clockwise is considered negative. In Fig. 1-3, $\angle RQP$ is positive while $\angle PQR$ is negative. These notations can also stand for the measures of angles, expressed either in degrees or in radians.

If we make an approximate guess as to the measures of the angles in Fig. 1-3, we might say that $\angle RQP = +60°$ while $\angle PQR = -60°$.

Rotational sense is not important in basic geometry, but it does matter when we work in coordinate geometry. We'll get into that type of geometry, which is also called *analytic geometry*, later in this book. For now, let's not worry about the rotational sense in which an angle is measured; we can consider all angles positive.

ANGLE BISECTION

Suppose there is an angle $\angle PQR$ measuring less than 180° and defined by three points P, Q, and R, as shown in Fig. 1-4. Then there is exactly one ray M that bisects (divides in half) the angle $\angle PQR$. If S is any point on M other than the point Q, then $\angle PQS = \angle SQR$. That is to say, every angle has one, and only one, ray that bisects it.

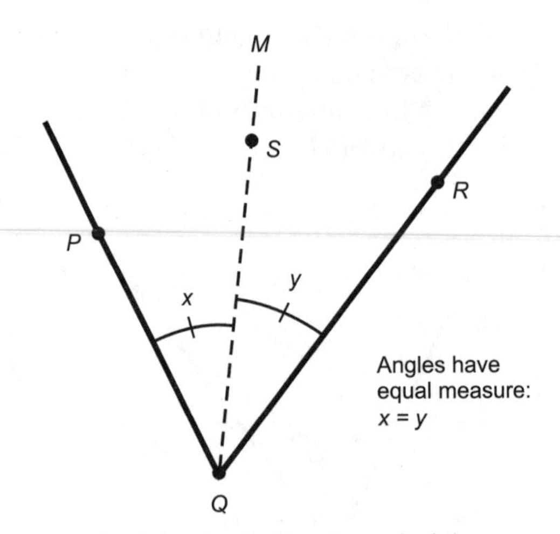

Fig. 1-4. Angle bisection principle.

PERPENDICULARITY

Suppose that L is a line through points P and Q. Let R be a point not on L. Then there is exactly one line M through point R, intersecting line L at some point S, such that M is perpendicular to L (that is, such that M and L intersect at a right angle). This is shown in Fig. 1-5. The term *orthogonal* is sometimes used instead of perpendicular. Another synonym for perpendicular, used especially in theoretical physics, is *normal*.

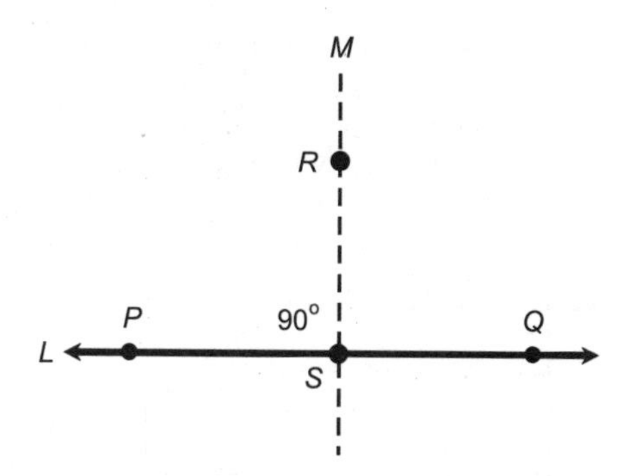

Fig. 1-5. Perpendicular principle.

PERPENDICULAR BISECTOR

Suppose that L is a line segment connecting two points P and R. Then there is one and only one line M that is perpendicular to L and that intersects L at a point Q, such that the distance from P to Q is equal to the distance from Q to R. That is, every line segment has exactly one *perpendicular bisector*. This is illustrated in Fig. 1-6.

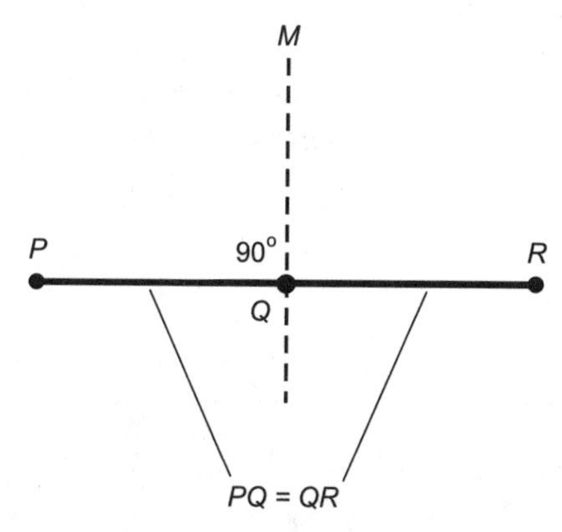

Fig. 1-6. Perpendicular bisector principle. Line M is unique.

DISTANCE ADDITION AND SUBTRACTION

Let P, Q, and R be points on a line L, such that Q is between P and R. Then the following equations hold concerning distances as measured along L (Fig. 1-7):

$$PQ + QR = PR$$
$$PR - PQ = QR$$
$$PR - QR = PQ$$

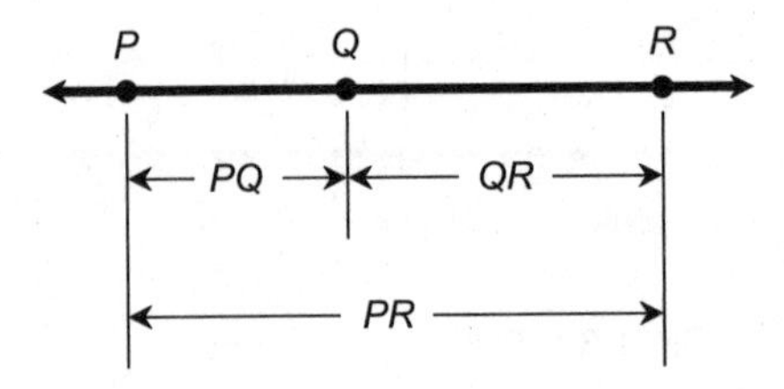

Fig. 1-7. Distance addition and subtraction.

ANGLE ADDITION AND SUBTRACTION

Suppose that P, Q, R, and S are points that all lie in the same plane. That is, they are all on a common, perfectly flat surface. Let Q be the vertex of three angles $\angle PQR$, $\angle PQS$, and $\angle SQR$, with ray QS between rays QP and QR as shown in Fig. 1-8. Then the following equations hold concerning the angular measures:

$$\angle PQS + \angle SQR = \angle PQR$$
$$\angle PQR - \angle PQS = \angle SQR$$
$$\angle PQR - \angle SQR = \angle PQS$$

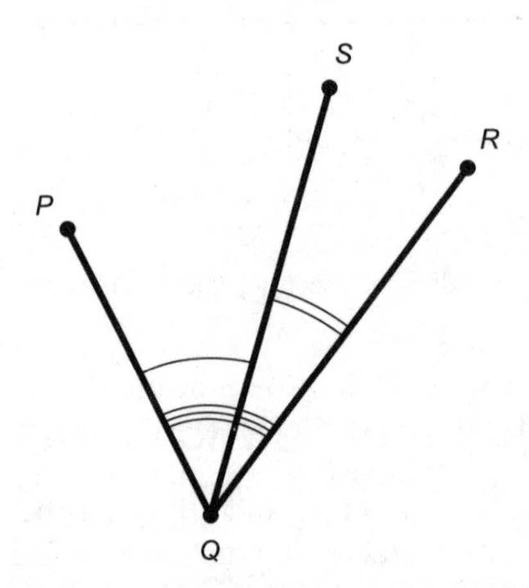

Fig. 1-8. Angular addition and subtraction.

PROBLEM 1-3

Look at Fig. 1-6. Suppose S is some point on line M other than point Q. What can we say about the lengths of line segments PS and SR?

SOLUTION 1-3

The solutions to problems like this can be made easier by making your own drawings. The more complicated the language (geometry problems can sometimes read like "legalese"), the more helpful drawings become. With the aid of your own sketch, you should be able to see that for every point S on line M (other than point Q, of course), the distances PS and SR are greater than the distances PQ and QR, respectively.

PROBLEM 1-4

Look at Fig. 1-8. Suppose that point S is moved perpendicularly with respect to the page (either straight toward you or straight away from you), so S no longer lies in the same plane as points P, Q, and R. What can we say about the measures of $\angle PQR$, $\angle PQS$, and $\angle SQR$?

SOLUTION 1-4

In this situation, the sum of the measures of $\angle PQS$ and $\angle SQR$ is greater than the measure of $\angle PQR$. This is because the measures of both $\angle PQS$ and $\angle SQR$ increase if point S departs perpendicularly from the plane containing points P, Q, and R. As point S moves further and further toward or away from you, the measures of $\angle PQS$ and $\angle SQR$ increase more and more.

More about Lines and Angles

In the confines of a single geometric plane, lines and angles behave according to various rules. The following are some of the best-known principles.

PARALLEL LINES

Two lines are *parallel* if and only if they lie in the same plane and they do not intersect at any point. Two line segments or rays that lie in the same plane are parallel if and only if, when extended infinitely in both directions to form complete lines, those complete lines do not intersect at any point.

COMPLEMENTARY AND SUPPLEMENTARY

Two angles that lie in the same plane are said to be *complementary angles* (they "complement" each other) if and only if the sum of their measures is 90° ($\pi/2$ rad). Two angles in the same plane are said to be *supplementary angles* (they "supplement" each other) if and only if the sum of their measures is 180° (π rad).

ADJACENT ANGLES

Suppose that L and M are two lines that intersect at a point P. Then any two *adjacent angles* between lines L and M are supplementary. This can be illustrated by drawing two intersecting lines, and noting that pairs of adjacent angles always form a *straight angle*, that is, an angle of 180° (π rad) determined by the intersection point and one of the two lines.

VERTICAL ANGLES

Suppose that L and M are two lines that intersect at a point P. Opposing pairs of angles, denoted x and y in Fig. 1-9, are known as *vertical angles.* Pairs of vertical angles always have equal measure. (The term "vertical" in this context is misleading; a better term would be "opposite" or "opposing." But a long time ago, somebody decided that "vertical" was good enough, and the term stuck.)

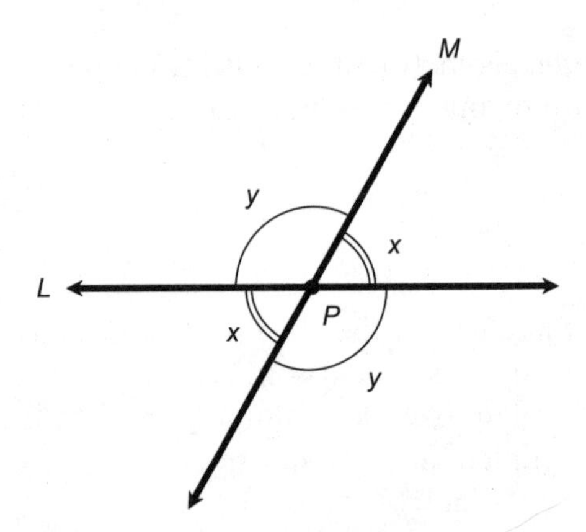

Fig. 1-9. Vertical angles have equal measure.

ALTERNATE INTERIOR ANGLES

Suppose that L and M are parallel lines. Let N be a line that intersects lines L and M at points P and Q, respectively. Line N is called a *transversal* to the parallel lines L and M. In Fig. 1-10, angles labeled x are *alternate interior angles*; the same holds true for angles labeled y. Pairs of alternate interior angles always have equal measure.

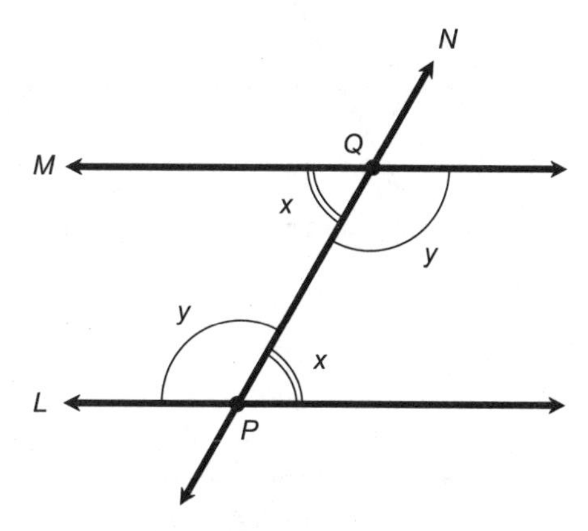

Fig. 1-10. Alternate interior angles have equal measure.

If line N is perpendicular to lines L and M, then $x = y$. Conversely, if $x = y$, then N is perpendicular to lines L and M. When a logical statement works both ways like this, the expression "if and only if" (often abbreviated "iff") is used. Here, $x = y$ iff N is perpendicular to both L and M. The phrase "is perpendicular to" is often replaced by the symbol $\perp$. So in shorthand, we can write ($N \perp L$ and $N \perp M$) iff $x = y$.

ALTERNATE EXTERIOR ANGLES

Suppose that L and M are parallel lines. Let N be a line that intersects L and M at points P and Q, respectively. In Fig. 1-11, angles labeled x are *alternate exterior angles*; the same holds true for angles labeled y. Pairs of alternate exterior angles always have equal measure. In addition, ($N \perp L$ and $N \perp M$) iff $x = y$.

CORRESPONDING ANGLES

Suppose that L and M are parallel lines. Let N be a line that intersects L and M at points P and Q, respectively. In Fig. 1-12, angles labeled w are *corre-*

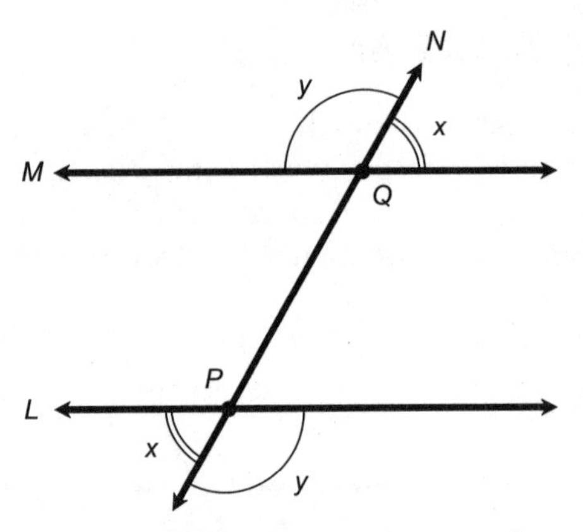

Fig. 1-11. Alternate exterior angles have equal measure.

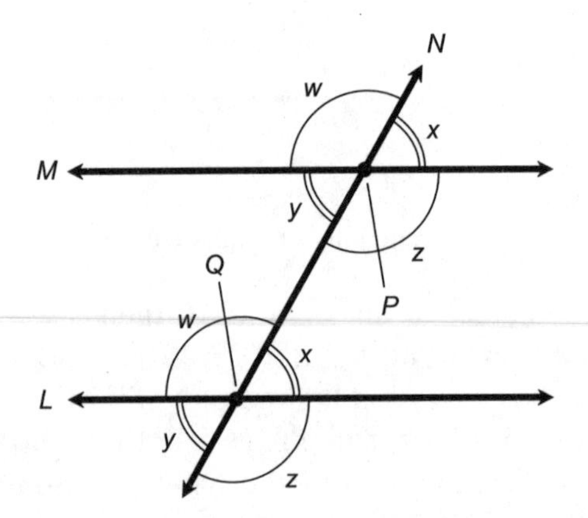

Fig. 1-12. Corresponding angles have equal measure.

sponding angles; the same holds true for angles labeled x, y, and z. Pairs of corresponding angles always have equal measure. In addition, N is perpendicular to both L and M if and only if one of the following is true:

$$w = x$$

$$y = z$$

$$w = y$$

$$x = z$$

In shorthand, this statement is written as follows:

$$(N \perp L \text{ and } N \perp M) \text{ iff } (w = x \text{ or } y = z \text{ or } w = y \text{ or } x = z)$$

PARALLEL PRINCIPLE

Suppose L is a line and P is a point not on L. Then there is one, but only one, line M through P, such that M is parallel to L (Fig. 1-13). This is known as the *parallel principle* or *parallel postulate*, and is one of the most important postulates in Euclidean geometry.

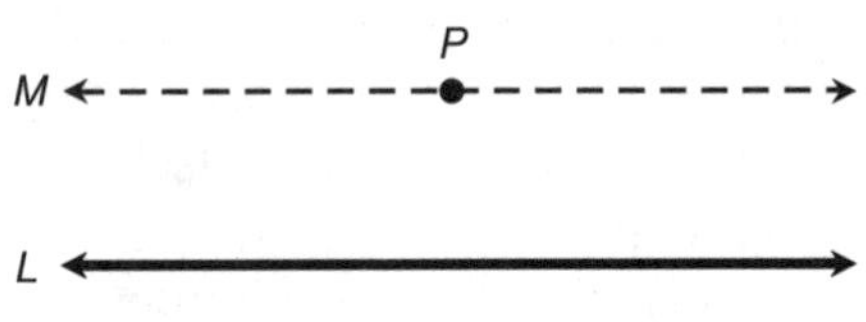

Fig. 1-13. The parallel principle.

In certain variants of geometry, the parallel postulate does not necessarily hold true. The denial of the parallel postulate forms the cornerstone of non-Euclidean geometry. We will look at this subject in Chapter 11.

PERPENDICULARITY REPEATED

Let L and M be lines that lie in the same plane. Suppose both L and M intersect a third line N, and both L and M are perpendicular to N. Then lines L and M are parallel to each other (Fig. 1-14).

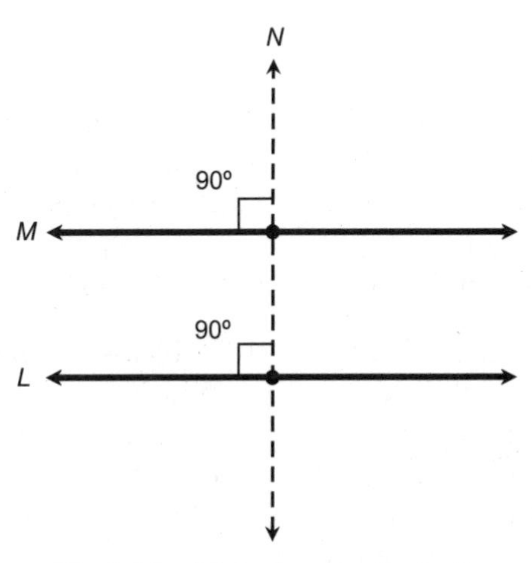

Fig. 1-14. Mutual perpendicularity.

In this drawing, the fact that two lines are perpendicular (they intersect at a right angle) is indicated by marking the intersection points with little "square-offs." This is a standard notation for indicating that lines, line segments, or rays are perpendicular at a point of intersection. Alternatively, we can write "90°" or "$\pi/2$ rad" near the intersection point.

PROBLEM 1-5

Suppose you are standing on the edge of a highway. The road is perfectly straight and flat, and the pavement is 20 meters wide everywhere. Suppose you lay a string across the road so it intersects one edge of the pavement at a 70° angle, measured with respect to the edge itself. If you stretch the string out so it is perfectly straight, and spool out enough of it so it crosses the other edge of the road, at what angle will the string intersect the other edge of the pavement, measured relative to that edge? At what angle will the string intersect the center line of the road, measured relative to the center line?

SOLUTION 1-5

This problem involves a double case of alternate interior angles, illustrated in Fig. 1-10. Alternatively, the principle for corresponding angles (Fig. 1-12) can be invoked. The edges of the pavement are parallel to each other, and also are both parallel to the center line. Therefore, the string will intersect the other edge of the road at a 70° angle; it will also cross the center line at a 70° angle. Note that these angles are expressed between the string and the pavement edges and center line themselves, not with respect to normals to the pavement edge or the center line (as is often done in physics).

PROBLEM 1-6

What are the measures of the above angles with respect to normals to the pavement edges and center line?

SOLUTION 1-6

A normal to any line always subtends an angle of 90° relative to that line. Thus, the string will cross both edges of the pavement at an angle of 90° −70°, or 20°, relative to the normal. We know this from the principle of angle addition and subtraction, shown in Fig. 1-8. The string will also cross the center line at an angle of 20° with respect to the normal.

Don't conduct experiments like those of Problems 1-5 and 1-6 on real roads. If you want to illustrate these things for yourself, make your "highway" with a long length of freezer paper, and perform the experiment in your home with the aid of a protractor, some string, a yardstick or meter stick, and a pencil. Don't let small children or animals trip or slip on the freezer paper, try to eat it, or otherwise have an accident with it.

Quiz

Refer to the text in this chapter if necessary. A good score is eight correct. Answers are in the back of the book.

1. An angle measures 30°. How many radians is this, approximately? You can use a calculator if you need it.
   (a) 0.3333 rad
   (b) 0.5000 rad
   (c) 0.5236 rad
   (d) 0.7854 rad

2. Consider a half-open line segment PQ, which includes the end point P but not the end point Q. Let line L_1 be the perpendicular bisector of PQ, and suppose that L_1 intersects the line segment PQ at point Q_1. Now imagine the half-open line segment PQ_1, which includes point P but not point Q_1. Let line L_2 be the perpendicular bisector of PQ_1, and suppose that L_2 intersects the line segment PQ_1 at point Q_2. Imagine this process being repeated, forming perpendicular bisectors L_3, L_4, L_5, ..., crossing line segment PQ at points Q_3, Q_4, Q_5, ..., which keep getting closer and closer to P. After how many repetitions of this process will the perpendicular bisector pass through point P? Draw a picture of this situation if you cannot envision it from this wording.
   (a) The perpendicular bisector will never pass through P, no matter how many times the process is repeated
   (b) The question cannot be answered without more information
   (c) This question is meaningless, because a half-open line segment cannot have a perpendicular bisector
   (d) This question is meaningless, because a half-open line segment has infinitely many perpendicular bisectors

3. Suppose that a straight section of railroad crosses a straight stretch of highway. The acute angle between the tracks and the highway center line measures exactly 1 rad. What is the measure of the obtuse angle between the tracks and the highway center line?
   (a) This question cannot be answered without more information
   (b) 1 rad
   (c) $\pi/2$ rad
   (d) $\pi - 1$ rad

4. An open line segment
   (a) contains neither of its end points
   (b) contains one of its end points
   (c) contains two of its end points
   (d) contains three of its end points

5. Two different, straight lines in a Euclidean plane are parallel if and only if
   (a) they intersect at an angle of π rad
   (b) they intersect at an angle of 2π rad
   (c) they intersect at one and only one point
   (d) they do not intersect at any point

6. Suppose you choose two points at random in a plane. How many Euclidean line segments exist that connect these two points?
   (a) None
   (b) One
   (c) More than one
   (d) Infinitely many

7. The measures of vertical angles between intersecting lines
   (a) always add up to 90°
   (b) always add up to 180°
   (c) always add up to 360°
   (d) depend on the angle at which the lines intersect

8. Two lines are orthogonal. The measure of the angle between them is therefore
   (a) 0°
   (b) π rad
   (c) 2π rad
   (d) $\pi/2$ rad

9. When an angle is bisected, two smaller angles are formed. These smaller angles
   (a) are obtuse
   (b) measure 90°
   (c) have equal measure
   (d) have measures that add up to 180°

10. Suppose two straight lines cross at a point P, and the lines are not perpendicular. Call the measures of the obtuse vertical angles x_1 and x_2. Which of the following equations is true?

(a) $x_1 < x_2$ (that is, x_1 is smaller than x_2)
(b) $x_1 > x_2$ (that is, x_1 is greater than x_2)
(c) $x_1 = x_2$
(d) $x_1 + x_2 = 180°$

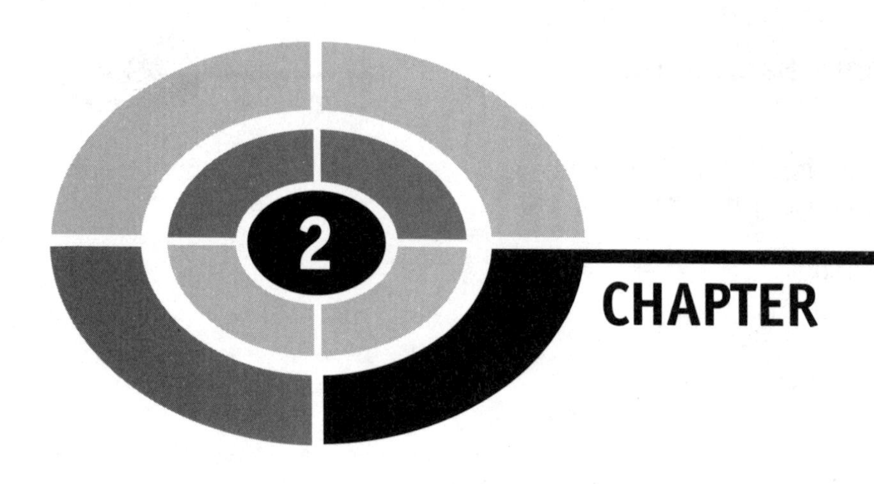

Triangles

If you ever took a course in plane geometry, you remember triangles. Do you recall being forced to learn formal proofs about them? We won't go through proofs here, but important facts about triangles are worth stating. If this is the first time you've worked with triangles, you should find most of the information in this chapter intuitively easy to grasp.

Triangle Definitions

In mathematics, it's essential to know exactly what one is talking about, without any "loopholes" or ambiguities. This is why there are formal definitions for almost everything (except primitives such as the point and the line).

WHAT IS A TRIANGLE?

First, let's define what a triangle is, so we will not make the mistake of calling something a triangle when it really isn't. A triangle is a set of three line segments, joined pairwise at their end points, and including those end points. The three points must not be collinear; that is, they must not all lie on the same straight line. For our purposes, we assume that the universe in which we define the triangle is Euclidean (not "warped" like the space around a black hole). In such an ideal universe, the shortest distance between any two points is defined by the straight line segment connecting those two points.

VERTICES

Figure 2-1 shows three points, called A, B, and C, connected by line segments to form a triangle. The points are called the *vertices* of the triangle. Often, other uppercase letters are used to denote the vertices of a triangle. For example, P, Q, and R are common choices.

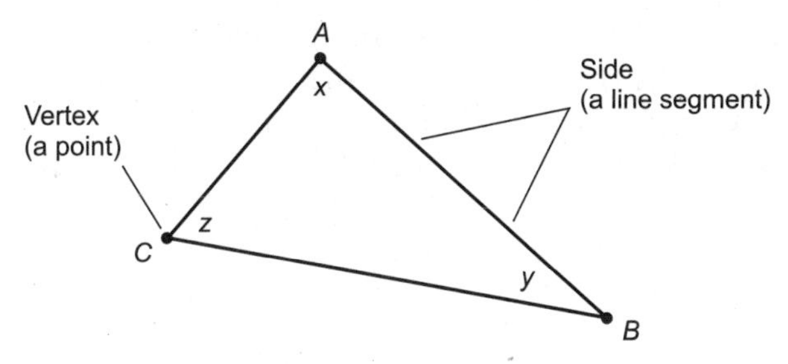

Fig. 2-1. Vertices, sides, and angles of a triangle.

NAMING

The triangle in Fig. 2-1 can be called, as you might guess, "triangle ABC." In geometry, it is customary to use a little triangle symbol (Δ) in place of the word "triangle." This symbol is actually the uppercase Greek letter delta. Fig. 2-1 illustrates a triangle that we can call ΔABC.

SIDES

The sides of the triangle in Fig. 2-1 are named according to their end points. Thus, ΔABC has three sides: line segment AB, line segment BC, and line segment CA. There are other ways of naming the sides, but as long as there is no confusion, we can call them just about anything.

INTERIOR ANGLES

Each vertex of a triangle is associated with an *interior angle*, which always measures more than 0° (0 rad) but less than 180° (π rad). In Fig. 2-1, the interior angles are denoted x, y, and z. Sometimes, italic lowercase Greek letters are used instead. Theta (pronounced "THAY-tuh") is a popular choice. It looks like a leaning numeral zero with a dash across it (θ). Subscripts can be used to denote the interior angles of a triangle, for example, θ_a, θ_b, and θ_c for the interior angles at vertices A, B, and C, respectively.

SIMILAR TRIANGLES

Two triangles are *directly similar* if and only if they have the same proportions in the same rotational sense. This means that one triangle is an enlarged and/or rotated copy of the other. Some examples of similar triangles are shown in Fig. 2-2. If you take any one of the triangles, enlarge it or reduce it uniformly and rotate it clockwise or counterclockwise to the correct extent, you can place it exactly over any of the other triangles. Two triangles are not directly similar if it is necessary to flip one of the triangles over, in addition to changing its size and rotating it, in order to be able to place it over the other.

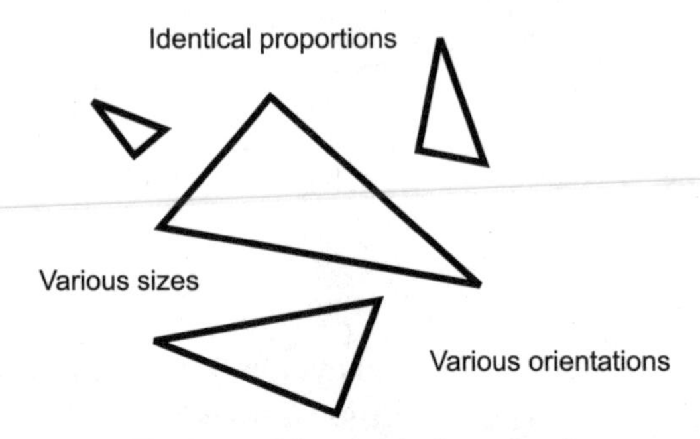

Fig. 2-2. Directly similar triangles.

Two triangles are *inversely similar* if and only if they are directly similar when considered in the opposite rotational sense. In simpler terms, they are inversely similar if and only if the mirror image of one is directly similar to the other.

If there are two triangles $\triangle ABC$ and $\triangle DEF$ that are directly similar, we can symbolize this by writing $\triangle ABC \sim \triangle DEF$. The direct similarity symbol looks like a wavy minus sign. If the triangles $\triangle ABC$ and $\triangle DEF$ are inversely similar, the situation is more complicated because there are three ways this can happen. Here they are:

- Points D and E are transposed, so $\triangle ABC \sim \triangle EDF$
- Points E and F are transposed, so $\triangle ABC \sim \triangle DFE$
- Points D and F are transposed, so $\triangle ABC \sim \triangle FED$

CONGRUENT TRIANGLES

There is disagreement in the literature about the exact meaning of the terms *congruence* and *congruent* when describing geometric figures in a plane. Some texts say two objects in a plane are congruent if and only if one can be placed exactly over the other after a rigid transformation (rotating it or moving it around, but not flipping it over). Other texts define congruence to allow flipping over, as well as rotation and motion. Let's stay away from that confusion, and make two definitions.

Two triangles exhibit *direct congruence* (they are *directly congruent*) if and only if they are directly similar, and the corresponding sides have the same lengths. Some examples are shown in Fig. 2-3. If you take one of the triangles and rotate it clockwise or counterclockwise to the correct extent, you can "paste" it precisely over any of the other triangles. Rotation and motion are allowed, but flipping over, also called *mirroring*, is forbidden. In general, triangles are not directly congruent if you must flip one of them over, in addition to rotating it, in order to be able to place it over the other.

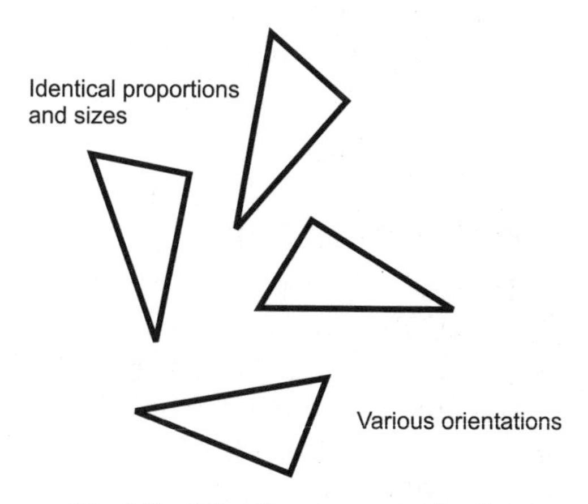

Fig. 2-3. Directly congruent triangles.

Two triangles exhibit *inverse congruence* (they are *inversely congruent*) if and only if they are inversely similar, and they are also the same size. Rotation and motion are allowed, and mirroring is actually required.

If there are two triangles $\triangle ABC$ and $\triangle DEF$ that are directly congruent, we can symbolize this by writing $\triangle ABC \cong \triangle DEF$. The direct congruence symbol is an equals sign with a direct similarity symbol on top. If the triangles $\triangle ABC$ and $\triangle DEF$ are inversely congruent, the same situation arises as is the case with inverse similarity. Three possibilities exist:

- Points D and E are transposed, so $\triangle ABC \cong \triangle EDF$
- Points E and F are transposed, so $\triangle ABC \cong \triangle DFE$
- Points D and F are transposed, so $\triangle ABC \cong \triangle FED$

TWO CRUCIAL FACTS

Here are two important things you should remember about triangles that are directly congruent.

If two triangles are directly congruent, then their corresponding sides have equal lengths as you proceed around both triangles in the same direction. The converse of this is also true. If two triangles have corresponding sides with equal lengths as you proceed around them both in the same direction, then the two triangles are directly congruent.

If two triangles are directly congruent, then their corresponding interior angles (that is, the interior angles opposite the corresponding sides) have equal measures as you proceed around both triangles in the same direction. The converse of this is not necessarily true. It is possible for two triangles to have corresponding interior angles with equal measures when you proceed around them both in the same direction, and yet the two triangles are not directly congruent.

TWO MORE CRUCIAL FACTS

Here are two "mirror images" of the facts just stated. They concern triangles that are inversely congruent. The wording is almost (but not quite) the same!

If two triangles are inversely congruent, then their corresponding sides have equal lengths as you proceed around the triangles in opposite directions. The converse of this is also true. If two triangles have corresponding sides with equal lengths as you proceed around them in opposite directions, then the two triangles are inversely congruent.

If two triangles are inversely congruent, then their corresponding interior angles have equal measures as you proceed around the triangles in opposite directions. The converse of this is not necessarily true. It is possible for two triangles to have corresponding interior angles with equal measures as you proceed around them in opposite directions, and yet the two triangles are not inversely congruent.

POINT–POINT–POINT PRINCIPLE

Let P, Q, and R be three distinct points that do not all lie on the same straight line. Then the following statements are true (Fig. 2-4):

- P, Q, and R lie at the vertices of some triangle T
- T is the only triangle having vertices P, Q, and R

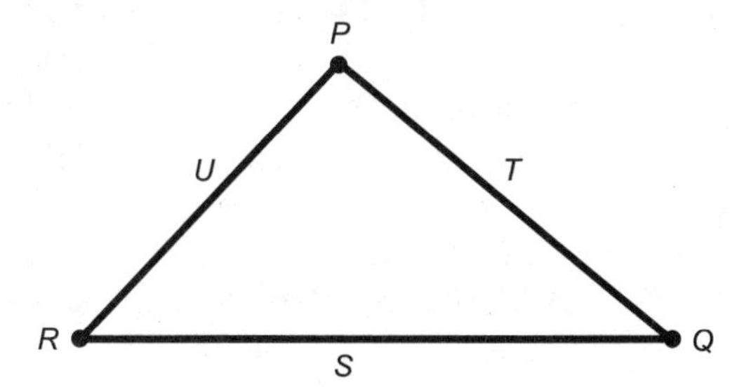

Fig. 2-4. The three-point principle; side–side–side triangles.

PROBLEM 2-1

Suppose you have a perfectly rectangular field surrounded by four straight lengths of fence. You build a straight fence diagonally across this field, so the diagonal fence divides the field into two triangles. Are these triangles directly congruent? If they are not congruent, are they directly similar?

SOLUTION 2-1

It helps to draw a diagram of this situation. If you do this, you can see that the two triangles are directly congruent. Consider the theoretical images of the triangles (which, unlike the fences, you can move around in your imagination). You can rotate one of these theoretical triangles exactly 180° (π rad), either clockwise or counterclockwise, and move it a short distance upward and to the side, and it will fit exactly over the other one.

PROBLEM 2-2

Suppose you have a telescope equipped with a camera. You focus on a distant, triangular sign and take a photograph of it. Then you double the magnification of the telescope and, making sure the whole sign fits into the field of view of the camera, you take another photograph. When you get the photos developed, you see triangles in each photograph. Are these triangles directly congruent? If not, are they directly similar?

SOLUTION 2-2

In the photos, one triangle looks larger than the other. But unless there is something wrong with the telescope, or you use a star diagonal when taking one photograph and not when taking the other (a star diagonal renders an image laterally inverted), the two triangle images have the same shape in the same rotational sense. They are not directly congruent, but they are directly similar.

Direct Congruence and Similarity Criteria

There are four criteria that can be used to define sets of triangles that are directly congruent. These are called the *side–side–side* (SSS), *side–angle–side* (SAS), *angle–side–angle* (ASA), and *angle–angle–side* (AAS) principles. The last of these can also be called *side–angle–angle* (SAA). A fifth principle, called *angle–angle–angle* (AAA), can be used to define sets of triangles that are directly similar.

SIDE–SIDE–SIDE (SSS)

Let S, T, and U be defined, specific line segments. Let s, t, and u be the lengths of those three line segments, respectively. Suppose that S, T, and U are joined at their end points P, Q, and R (Fig. 2-4). Then the following statements hold true:

- Line segments S, T, and U determine a triangle
- This is the only triangle that has sides S, T, and U in this order, as you proceed around the triangle in the same rotational sense
- All triangles having sides of lengths s, t, and u in this order, as you proceed around the triangles in the same rotational sense, are directly congruent

SIDE–ANGLE–SIDE (SAS)

Let S and T be two distinct line segments. Let P be a point that lies at the ends of both of these line segments. Denote the lengths of S and T by their lowercase counterparts s and t, respectively. Suppose S and T subtend an angle x, expressed in the counterclockwise sense, at point P (Fig. 2-5). Then the following statements are all true:

- S, T, and x determine a triangle
- This is the only triangle having sides S and T that subtend an angle x, measured counterclockwise from S to T, at point P

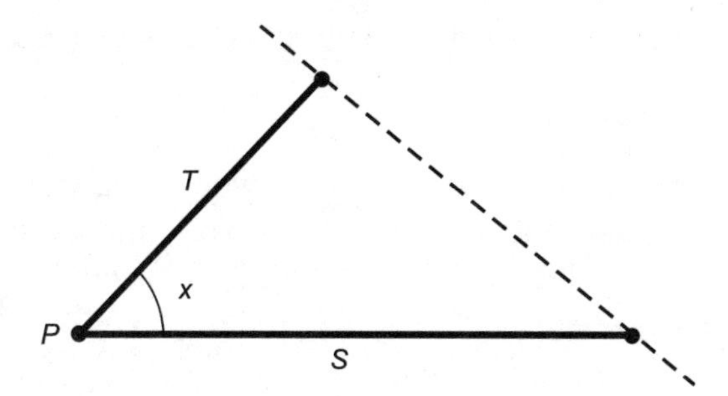

Fig. 2-5. Side–angle–side triangles.

- All triangles containing two sides of lengths s and t that subtend an angle x, measured counterclockwise from the side of length s to the side of length t, are directly congruent

ANGLE–SIDE–ANGLE (ASA)

Let S be a line segment having length s, and whose end points are P and Q. Let x and y be the angles subtended relative to S by two lines L and M that run through P and Q, respectively (Fig. 2-6), such that both angles are expressed in the counterclockwise sense. Then the following statements are all true:

- x, S, and y determine a triangle
- This is the only triangle determined by x, S, and y, proceeding from left to right
- All triangles containing one side of length s, and whose other two sides subtend angles of x and y relative to the side whose length is s, with x on the left and y on the right and both angles expressed in the counterclockwise sense, are directly congruent

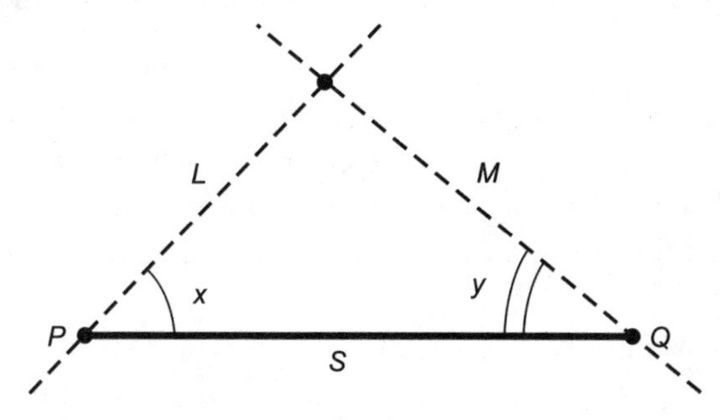

Fig. 2-6. Angle–side–angle triangles.

ANGLE–ANGLE–SIDE (AAS) OR SIDE–ANGLE–ANGLE (SAA)

Let S be a line segment having length s, and whose end points are P and Q. Let x and y be angles, one adjacent to S and one opposite, and both expressed in the counterclockwise sense (Fig. 2-7). Then the following statements are all true:

- S, x, and y determine a triangle
- This is the only triangle determined by S, x, and y in the counterclockwise sense
- All triangles containing one side of length s, and two angles x and y expressed and proceeding in the counterclockwise sense, are directly congruent

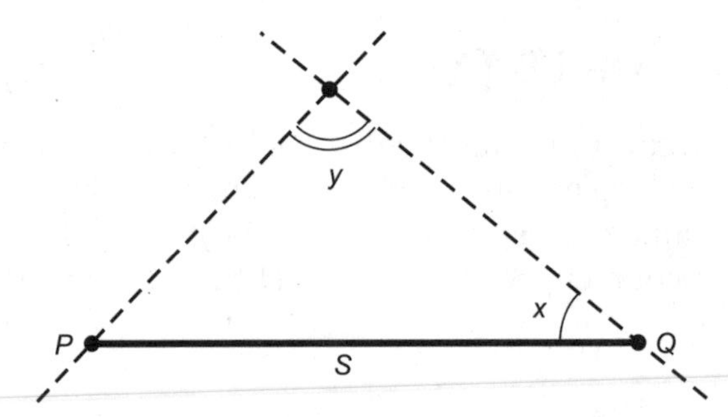

Fig. 2-7. Angle–angle–side triangles.

ANGLE–ANGLE–ANGLE (AAA)

Let L, M, and N be lines that lie in a common plane and intersect in three points as illustrated in Fig. 2-8. Let the angles at these points, all expressed in the counterclockwise sense, be x, y, and z. Then the following statements are all true:

- There are infinitely many triangles with interior angles x, y, and z, in this order and proceeding in the counterclockwise sense
- All triangles with interior angles x, y, and z, in this order, expressed and proceeding in the counterclockwise sense, are directly similar

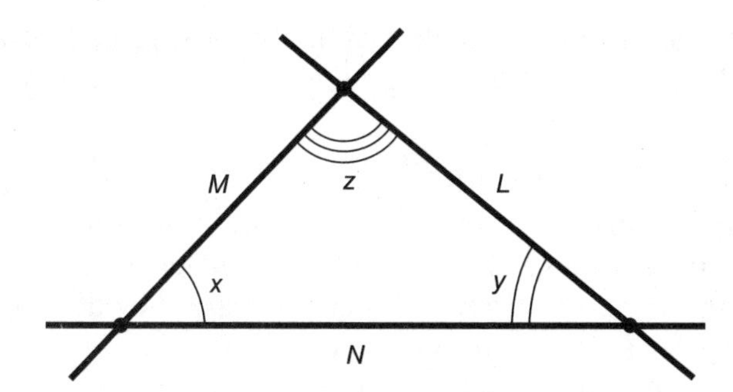

Fig. 2-8. Angle–angle–angle triangles.

LET IT BE SO!

Are you wondering why the word "let" is used so often? For example, "Let P, Q, and R be three distinct points." This sort of language is customary. You'll find it all the time in mathematical literature. When you are admonished to "let" things be a certain way, you are in effect being asked to imagine, or suppose, that things are such, to set the scene in your mind for statements or problems that follow.

PROBLEM 2-3

Refer to Fig. 2-6. Suppose x and y both measure 60°. If the resulting triangle is reversed from left to right—that is, flipped over around a vertical axis—will the resulting triangle be directly similar to the original? Will it be directly congruent to the original?

SOLUTION 2-3

This is a special case in which a triangle can be flipped over and the result is not only inversely congruent, but also directly congruent, to the original. This is the case because the triangle is symmetrical with respect to a straight-line axis. To clarify this, draw a triangle after the pattern in Fig. 2-6, but using a protractor to generate 60° angles for both x and y. (As it is drawn in this book, the figure is not symmetrical and the angles are not both 60°.) Then look at the image you have drawn, both directly and while standing in front of a mirror. The two mirror-image triangles are, in this particular case, identical.

PROBLEM 2-4

Suppose, in the situation of Problem 2-3, you split the triangle, whose angles x and y both measure 60°, right down the middle. You do this by dropping a vertical line from the top vertex so it intersects line segment PQ at its midpoint. Are the resulting two triangles, each comprising half of the original, directly similar? Are they directly congruent? Are they inversely similar? Are they inversely congruent?

SOLUTION 2-4

These triangles are mirror images of each other, but you cannot magnify, reduce, and/or rotate one of these triangles to make it fit exactly over the other. The triangles are not directly similar, nor are they directly congruent, even though, in a sense, they are the same size and shape.

Remember that for two triangles to be directly similar, the lengths of their sides must be in the same proportion, in order, as you proceed in the same rotational sense (counterclockwise or clockwise) around them both. In order to be directly congruent, their sides must have identical lengths, in order, as you proceed in the same rotational sense, around both.

These two triangles are inversely similar and inversely congruent, because they are mirror images of each other and are the same size.

Types of Triangles

Triangles can be categorized qualitatively (that means according to their qualities or characteristics). Here are the most common character profiles.

ACUTE TRIANGLE

When each of the three interior angles of a triangle are acute, that triangle is called an *acute triangle*. In such a triangle, none of the angles measures as much as 90° ($\pi/2$ rad). Examples of acute triangles are shown in Fig. 2-9.

Fig. 2-9. In an acute triangle, all angles measure less than 90° ($\pi/2$ rad).

OBTUSE TRIANGLE

When one of the interior angles of a triangle is obtuse, that triangle is called an *obtuse triangle*. Such a triangle has one obtuse interior angle, that is, one angle that measures more than 90° ($\pi/2$ rad). Some examples are shown in Fig. 2-10.

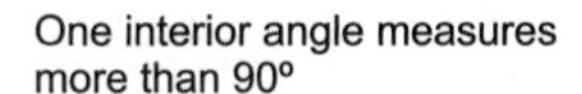

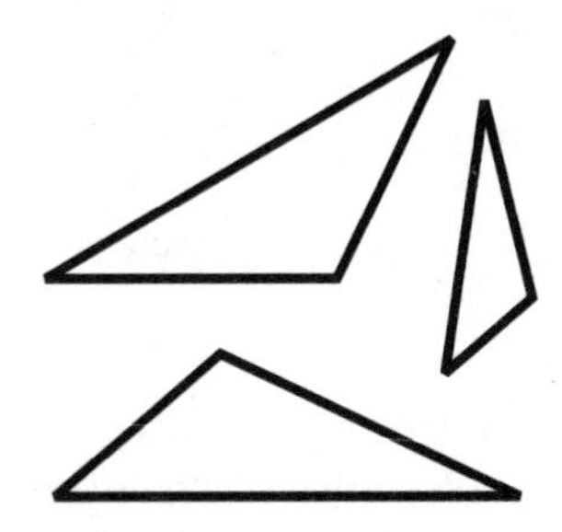

Fig. 2-10. In an obtuse triangle, one angle measures more than 90° ($\pi/2$ rad).

ISOSCELES TRIANGLE

Suppose we have a triangle with sides S, T, and U, having lengths s, t, and u, respectively. Let x, y, and z be the angles opposite sides S, T, and U, respectively. Suppose any of the following equations hold:

$$s = t$$
$$t = u$$
$$s = u$$
$$x = y$$
$$y = z$$
$$x = z$$

One example of such a situation is shown in Fig. 2-11. This kind of triangle is called an *isosceles triangle*, and the following logical statements are true:

$$s = t \Longleftrightarrow x = y$$
$$t = u \Longleftrightarrow y = z$$
$$s = u \Longleftrightarrow x = z$$

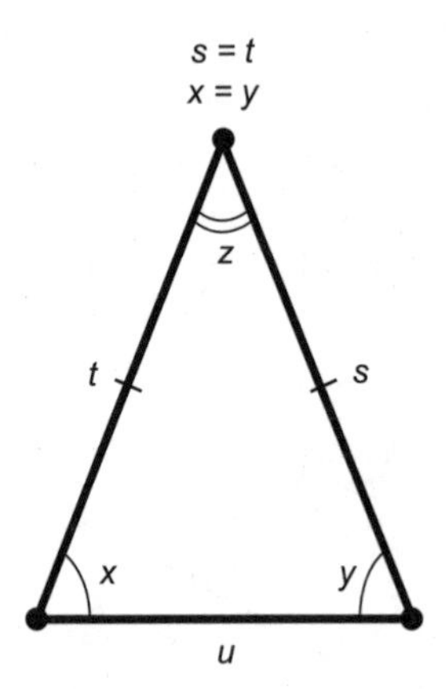

Fig. 2-11. Isosceles triangle.

The double-shafted double arrow ($\Longleftrightarrow$) means "if and only if." It is well to remember this. You should also know that a double-shafted single arrow pointing to the right ($\Rightarrow$) stands for "implies" or "means it is always true that." When we say $s = t \Longleftrightarrow x = y$, it is logically equivalent to saying $s = t \Rightarrow x = y$ and also $x = y \Rightarrow s = t$.

EQUILATERAL TRIANGLE

Suppose we have a triangle with sides S, T, and U, having lengths s, t, and u, respectively. Let x, y, and z be the angles opposite sides S, T, and U, respectively. Suppose either of the following are true:

$$s = t = u$$

or

$$x = y = z$$

Then the triangle is said to be an *equilateral triangle* (Fig. 2-12), and the following logical statement is valid:

$$s = t = u \Longleftrightarrow x = y = z$$

This means that all equilateral triangles have precisely the same shape; they are all directly similar. (They all happen to be inversely similar, too.)

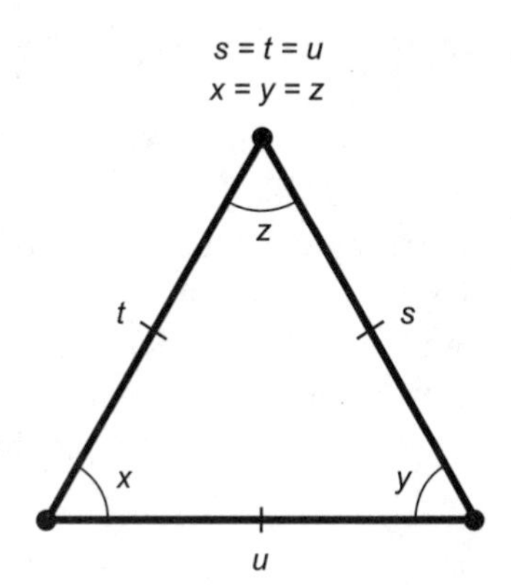

Fig. 2-12. Equilateral triangle.

RIGHT TRIANGLE

Suppose we have a triangle $\triangle PQR$ with sides S, T, and U, having lengths s, t, and u, respectively. If one of the interior angles of this triangle measures 90° ($\pi/2$ rad), an angle that is also called a *right angle*, then the triangle is called a *right triangle*. In Fig. 2-13, a right triangle is shown in which $\angle PRQ$ is a right angle. The side opposite the right angle is the longest side, and is called the *hypotenuse*. In Fig. 2-13, this is the side of length u.

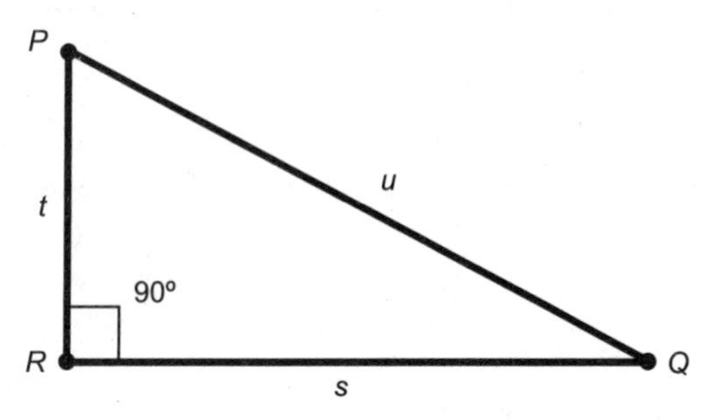

Fig. 2-13. Right angle

Special Facts

Triangles have some special properties. These characteristics have applications in many branches of science and engineering.

A TRIANGLE DETERMINES A UNIQUE PLANE

The vertex points of a specific triangle define one, and only one, Euclidean (that is, flat) geometric plane. A specific Euclidean plane can, however, contain infinitely many different triangles. This is intuitively obvious when you give it a little thought. Just try to imagine three points that don't all lie in the same plane! Incidentally, this principle explains why a three-legged stool never wobbles. It is the reason why cameras and telescopes are commonly mounted on tripods (three-legged structures) rather than structures with four or more legs.

SUM OF ANGLE MEASURES

In any triangle, the sum of the measures of the interior angles is 180° (π rad). This holds true regardless of whether it is an acute, right, or obtuse triangle, as long as all the angles are measured in the plane defined by the three vertices of the triangle.

THEOREM OF PYTHAGORAS

Suppose we have a right triangle defined by points P, Q, and R whose sides are S, T, and U having lengths s, t, and u, respectively. Let u be the hypotenuse (Fig. 2-13). Then the following equation is always true:

$$s^2 + t^2 = u^2$$

The converse of this is also true: If there is a triangle whose sides have lengths s, t, and u, and the above equation is true, then that triangle is a right triangle.

PERIMETER OF TRIANGLE

Suppose we have a triangle defined by points P, Q, and R, and having sides S, T, and U of lengths s, t, and u, as shown in Fig. 2-14. Then the perimeter, B, of the triangle is given by the following formula:

$$B = s + t + u$$

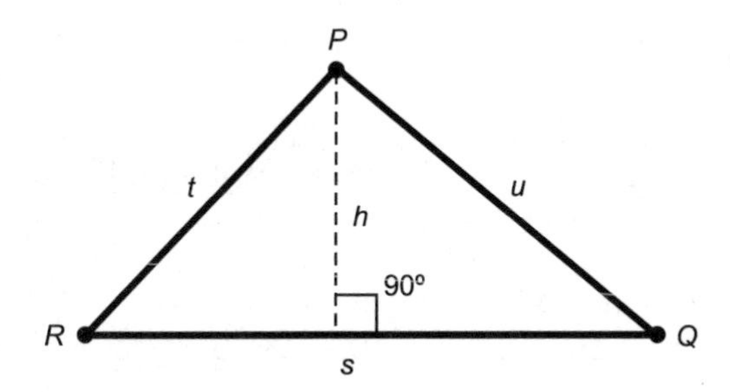

Fig. 2-14. Perimeter and area of triangle.

INTERIOR AREA OF TRIANGLE

Consider the same triangle as defined above; refer again to Fig. 2-14. Let s be the base length, and let h be the height, or the length of a perpendicular line segment between point P and side S. The interior area, A, can be found with this formula:

$$A = sh/2$$

PROBLEM 2-5

Suppose that $\triangle PQR$ in Fig. 2-14 has sides of lengths $s = 10$ meters, $t = 7$ meters, and $u = 8$ meters. What is the perimeter B of this triangle?

SOLUTION 2-5

Simply add up the lengths of the sides:

$$\begin{aligned} B &= s + t + u \\ &= (10 + 7 + 8) \text{ meters} \\ &= 25 \text{ meters} \end{aligned}$$

PROBLEM 2-6

Are there any triangles having sides of lengths 10 meters, 7 meters, and 8 meters, in that order proceeding clockwise, that are not directly congruent to $\triangle PQR$ as described in Problem 2-5?

SOLUTION 2-6

No. According to the side–side–side (SSS) principle, all triangles having sides of lengths 10 meters, 7 meters, and 8 meters, in this order as you proceed in the same rotational sense, are directly congruent.

Quiz

Refer to the text in this chapter if necessary. A good score is eight correct. Answers are in the back of the book.

1. Suppose there are three triangles, called $\triangle ABC$, $\triangle DEF$, and $\triangle PQR$. If $\triangle ABC \cong \triangle DEF$ and $\triangle DEF \cong \triangle PQR$, we can surmise that $\triangle ABC$ and $\triangle PQR$ are
   (a) directly congruent
   (b) directly similar, but not directly congruent
   (c) inversely congruent
   (d) not related in any particular way

2. Suppose there are two triangles, called $\triangle ABC$ and $\triangle DEF$. If these two triangles are directly similar, then we can be certain that
   (a) $\angle ABC$ and $\angle DFE$ have equal measure
   (b) $\angle BCA$ and $\angle EFD$ have equal measure
   (c) $\angle CAB$ and $\angle FED$ have equal measure
   (d) both triangles are equilateral

3. Suppose a given triangle is directly congruent to its mirror image. We can be absolutely certain that this triangle is
   (a) equilateral
   (b) isosceles
   (c) acute
   (d) obtuse

4. Suppose a triangle has sides of lengths s, t, and u, in centimeters (cm). Which of the following situations represents a right triangle? Assume the lengths are mathematically exact (no measurement error).
   (a) $s = 2$ cm, $t = 3$ cm, $u = 4$ cm
   (b) $s = 4$ cm, $t = 5$ cm, $u = 7$ cm
   (c) $s = 6$ cm, $t = 8$ cm, $u = 10$ cm
   (d) $s = 7$ cm, $t = 11$ cm, $u = 13$ cm

5. Suppose there are two triangles, called $\triangle ABC$ and $\triangle DEF$. Also suppose that side DE is twice as long as side AB, side EF is twice as long as side BC, and side DF is twice as long as side AC. Which of the following statements is true?
   (a) The interior area of $\triangle DEF$ is twice the interior area of $\triangle ABC$
   (b) The perimeter of $\triangle DEF$ is four times the perimeter of $\triangle ABC$

(c) The interior area of $\triangle DEF$ is four times the interior area of $\triangle ABC$
(d) $\triangle ABC \cong \triangle DEF$

6. Suppose a triangle has a base length of 4 feet and a height of 4 feet. Its interior area is
   (a) 4 square feet
   (b) 8 square feet
   (c) 16 square feet
   (d) impossible to determine without more information

7. The perimeter of the triangle described in the previous question is
   (a) 8 feet
   (b) 16 feet
   (c) 22 feet
   (d) impossible to determine without more information

8. Draw a triangle on a piece of paper. Label the vertices {, [, and (, proceeding in a counterclockwise sense. Look at this figure and call it △{[(. Now hold the piece of paper between your eyes and a bright light, and turn the inked side away from you but keep the page right-side-up. You should see a figure that you can call △)]}, because the vertices appear as),], and } in a counterclockwise sense. Which of the following statements is true?
   (a) These two triangles are directly congruent
   (b) These two triangles are directly similar, but not directly congruent
   (c) These two triangles are inversely congruent
   (d) These two triangles are inversely similar, but not inversely congruent

9. Take the same piece of paper that you used in the preceding problem. Look at △{[(. Now turn the paper upside down, keeping the inked side facing you. You should see another triangle that you can call △}]) as the vertices appear in a counterclockwise sense. Which of the following statements is true?
   (a) These two triangles are directly congruent
   (b) These two triangles are directly similar, but not directly congruent
   (c) These two triangles are inversely congruent
   (d) These two triangles are inversely similar, but not inversely congruent

10. Suppose there is a triangle, two of the interior angles of which measure $30°$. The measure of the third angle is
    (a) impossible to determine without more information
    (b) $\pi/6$ rad
    (c) $\pi/4$ rad
    (d) $2\pi/3$ rad

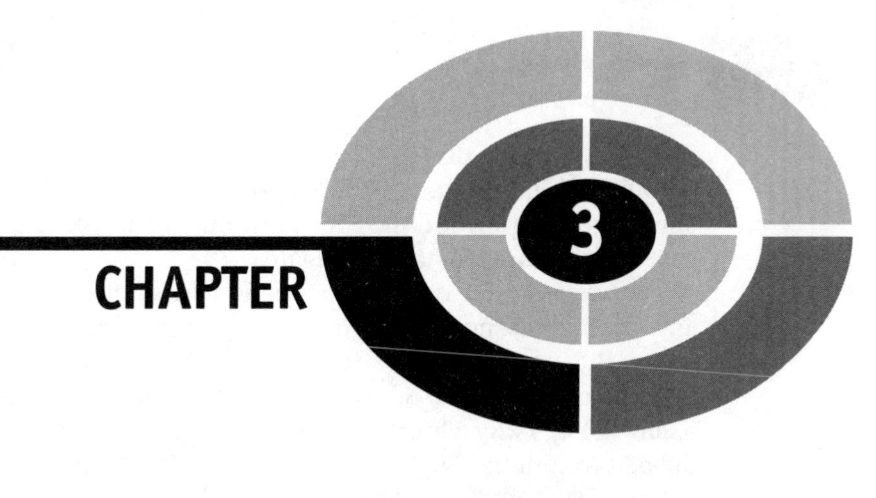

Quadrilaterals

A four-sided geometric plane figure is called a *quadrilateral*. Because a quadrilateral has more sides than a triangle, there are more types. The allowable range of interior-angle measures is greater than is the case with triangles. With a triangle, an interior angle must always measure more than $0°$ (0 rad) but less than $180°$ (π rad); with a quadrilateral, the measure of an interior angle can be anything up to, but not including, $360°$ (2π rad).

Types of Quadrilaterals

The categories of quadrilateral are the *square*, the *rhombus*, the *rectangle*, the *parallelogram*, the *trapezoid*, and the *general quadrilateral*. Let's define these and look at some examples.

IT'S THE LAW!

There are two properties that a four-sided geometric figure absolutely must have—laws it is required to obey—if it is to qualify as a legitimate plane quadrilateral. First, all four vertices must lie in the same plane. Second, all four sides must be straight line segments of finite, positive length. Curves are not allowed, nor are points, infinitely long rays, or infinitely long lines. For our purposes, we'll add the constraint that a true plane quadrilateral cannot have sides whose lengths are negative.

The vertices of a triangle always lie in a single geometric plane, because any three points, no matter which ones you choose, define a unique geometric plane. But when you have four points, they don't all necessarily lie in the same plane. Any three of them do, but the fourth one can get "out of alignment." This is why a four-legged stool or table often wobbles, and why it is so difficult to trim the lengths of the legs so the wobbling stops. Once the ends of the legs lie in a single plane, and they define the vertices of a plane quadrilateral, the stool or table won't wobble, as long as the floor is perfectly flat. (Later in this book, we'll take a look at some of the things that can happen when a floor is not flat, or more particularly, what can take place when a geometric universe is warped or curved.)

SQUARE

A square has four sides that are all of the same length. In addition, all the interior angles are the same, and measure 90° ($\pi/2$ rad). Figure 3-1 shows the general situation. The length of each side in this illustration is s units. There is no limit to how large s can be, but it must be greater than zero.

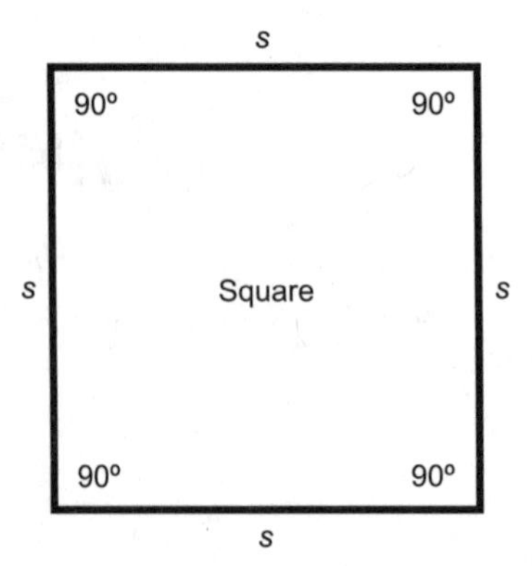

Fig. 3-1. Example of a square. Sides have length s, and the interior angles all measure 90° ($\pi/2$ rad).

RHOMBUS

A rhombus is like a square in that all four sides are the same length. But the angles don't all have to be right angles. A square is a special type of rhombus in which all four angles happen to have the same measure. But most rhombuses (rhombi?) look something like the example in Fig. 3-2. All four sides have length s. Opposite angles have equal measure, but adjacent angles need not. In this illustration, the two angles labeled x have equal measure, as do the two angles labeled y. Another property of the rhombus is the fact that both pairs of opposite sides are parallel.

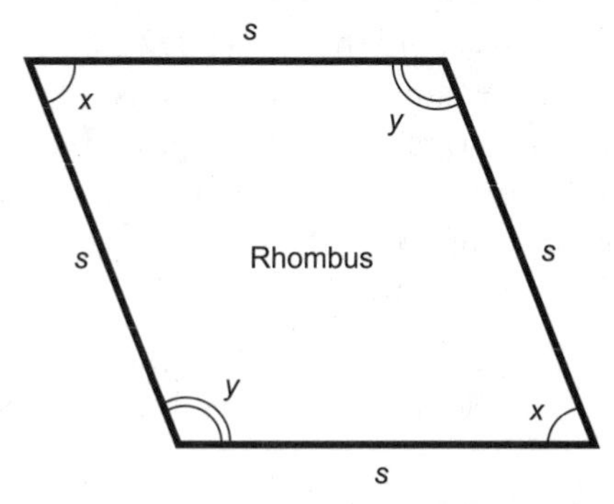

Fig. 3-2. Example of a rhombus. Sides have length s, while x and y denote interior angle measures.

RECTANGLE

A rectangle is like a square in that all four angles have equal measure. But the sides don't all have to be equally long. A square is a special type of rectangle in which all four sides happen to be the same length. But most rectangles look something like the example in Fig. 3-3. All four angles have the same measure, which must be 90° ($\pi/2$ rad). Opposite sides have equal length, but adjacent sides need not. In this illustration, the two sides labeled s have equal measure, as do the two sides labeled t.

PARALLELOGRAM

The defining characteristic of a parallelogram is that both pairs of opposite sides are parallel. This alone is sufficient to make a plane quadrilateral qualify as a parallelogram. Whenever both pairs of opposite sides in a quadrilateral

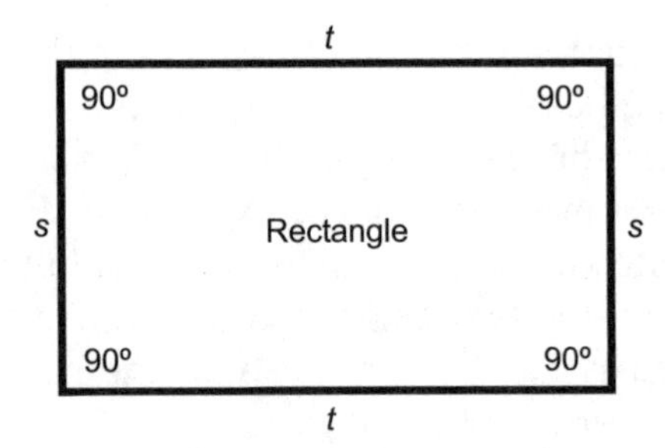

Fig. 3-3. Example of a rectangle. Sides have lengths s and t, while the interior angles all measure 90° ($\pi/2$ rad).

are parallel, those pairs also have the same length. In addition, pairs of opposite angles have equal measure. A rectangle is a special sort of parallelogram. So is a rhombus, and so is a square. Figure 3-4 shows an example of a parallelogram in which both angles labeled x have equal measure, both angles labeled y have equal measure, both sides labeled s are the same length, and both sides labeled t are the same length.

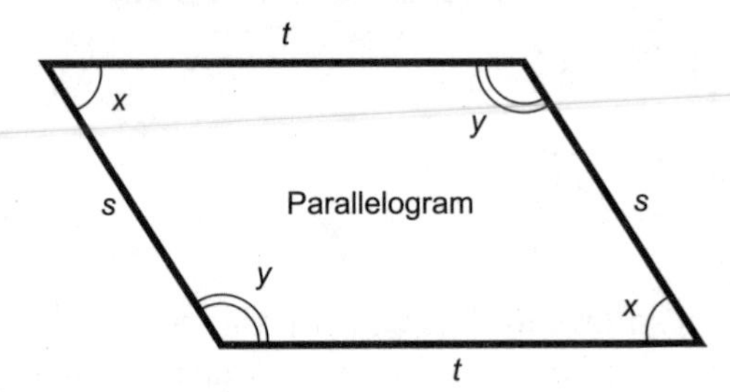

Fig. 3-4. Example of a parallelogram. Sides have lengths s and t, while x and y denote interior angle measures.

TRAPEZOID

If we remove yet another restriction from the quadrilateral, we get a trapezoid. The only rule a trapezoid must obey is that one pair of opposite sides must be parallel. Otherwise, anything goes! Figure 3-5 shows an example of a trapezoid. The dashed lines represent parallel lines in which the two parallel sides of the quadrilateral happen to lie. (The dashed lines are not part of the quadrilateral itself.)

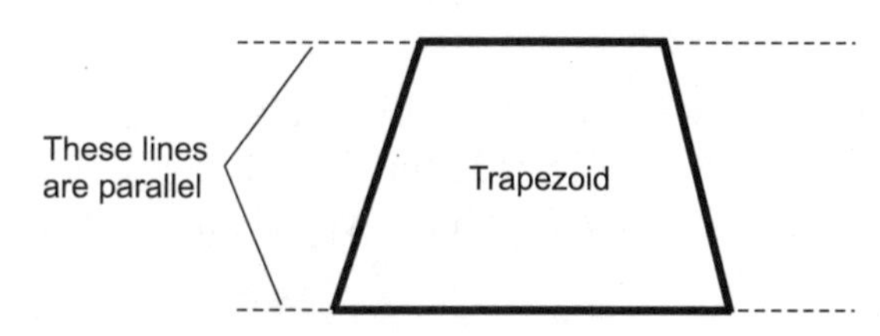

Fig. 3-5. In a trapezoid, one pair of opposite sides is parallel.

GENERAL QUADRILATERAL

In a general quadrilateral, there are no restrictions at all on the lengths of the sides, although the "nature of the beast" dictates that no angle can be outside the range 0° (0 rad) to 360° (2π rad), non-inclusive. As long as all four vertices lie in the same geometric plane, and as long as all four sides of the figure are straight line segments of finite and positive length, it's all right.

Of course, any quadrilateral can be considered "general." A rectangle, for example, is just a specific type of general quadrilateral. But there are plenty of general quadrilaterals that don't fall into any of the above categories. They don't exhibit any sort of symmetry or apparent orderliness. These are known as *irregular quadrilaterals*. Three examples are shown in Fig. 3-6.

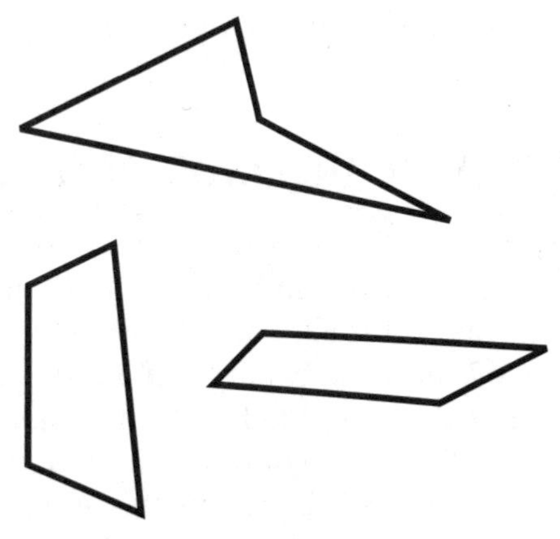

Fig. 3-6. Three examples of irregular quadrilaterals. The sides all have different lengths, and the angles all have different measures.

PROBLEM 3-1

What type of quadrilateral is formed by the boundaries of a soccer field?

SOLUTION 3-1

Assuming the groundskeepers have done their job correctly, a soccer field is shaped like a rectangle. All four corners form right angles (90°). In addition, the lengths of opposite sides are equal. The two sidelines are the same length, as are the two end lines.

PROBLEM 3-2

Suppose a quadrilateral *ABCD* is defined with the vertices going counterclockwise in alphabetical order. Suppose further that $\angle ABC = \angle CDA$ and $\angle BCD = \angle DAB$. What can be said about this quadrilateral?

SOLUTION 3-2

It helps to draw pictures here. Draw several examples of quadrilaterals that meet these two requirements. You'll see that $\angle ABC$ is opposite $\angle CDA$, and $\angle BCD$ is opposite $\angle DAB$. The fact that opposite pairs of angles have equal measure means that the quadrilateral must be a parallelogram. It might be a special case of the parallelogram, such as a rhombus, rectangle, or square; but the only restriction we are given is the fact that $\angle ABC = \angle CDA$ and $\angle BCD = \angle DAB$. Therefore, *ABCD* can be any sort of parallelogram.

Facts about Quadrilaterals

Every quadrilateral has certain properties, depending on the "species." Here are some useful facts concerning these four-sided plane figures.

SUM OF MEASURES OF INTERIOR ANGLES

No matter what the shape of a quadrilateral, as long as all four sides are straight line segments of positive and finite length, and as long as all four vertices lie in the same plane, the sum of the measures of the interior angles is 360° (2π rad). Figure 3-7 shows an example of an irregular quadrilateral. The interior angles are denoted w, x, y, and z. In this example, angle w measures more than 180° (π rad). If you use your imagination, you might call this type of quadrilateral a "boomerang," although this is not an official geometric term.

$$w + x + y + z = 360°$$

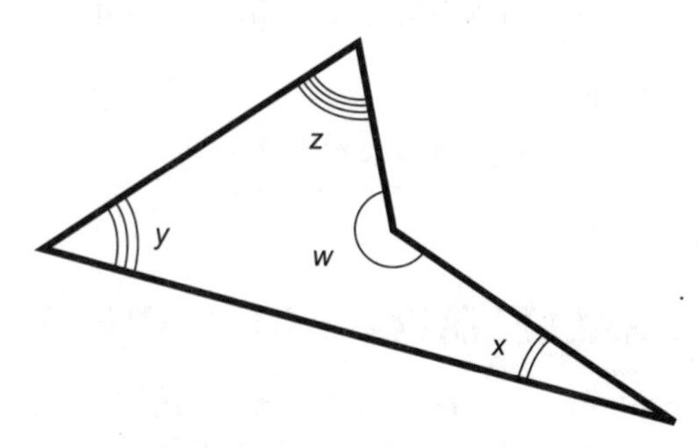

Fig. 3-7. In any plane quadrilateral, the sum of the measures of the interior angles w, x, y, and z is 360° (2π rad).

PARALLELOGRAM DIAGONALS

Suppose we have a parallelogram defined by four points P, Q, R, and S. Let D be a line segment connecting P and R as shown in Fig. 3-8A. Then D is a *minor diagonal* of the parallelogram, and the following triangles defined by D are congruent:

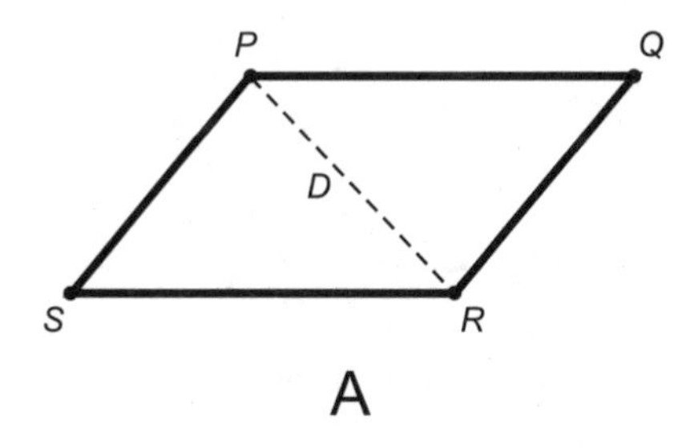

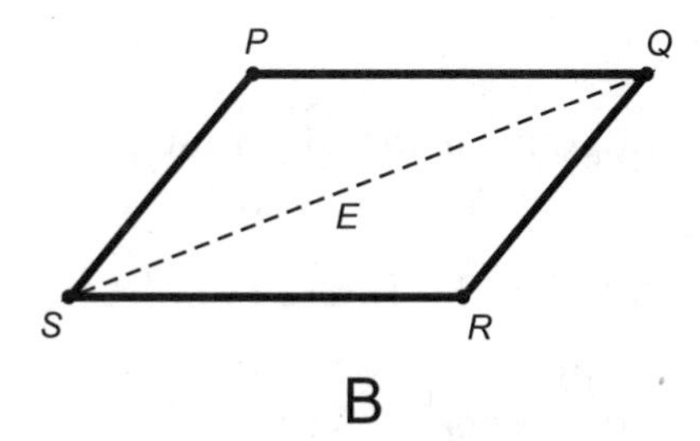

Fig. 3-8. Triangles defined by the minor diagonal (A) or the major diagonal (B) of a parallelogram are congruent.

Let E be a line segment connecting Q and S (Fig. 3-8B). Then E is a *major diagonal* of the parallelogram, and the following triangles defined by E are congruent:

$$\triangle QRS \cong \triangle SPQ$$

BISECTION OF PARALLELOGRAM DIAGONALS

Suppose we have a parallelogram defined by four points P, Q, R, and S. Let D be the diagonal connecting P and R; let E be the diagonal connecting Q and S (Fig. 3-9). Then D and E bisect each other at their intersection point T. In addition, the following pairs of triangles are congruent:

$$\triangle PQT \cong \triangle RST$$
$$\triangle QRT \cong \triangle SPT$$

The converse of the foregoing is also true: If we have a plane quadrilateral whose diagonals bisect each other, then that quadrilateral is a parallelogram.

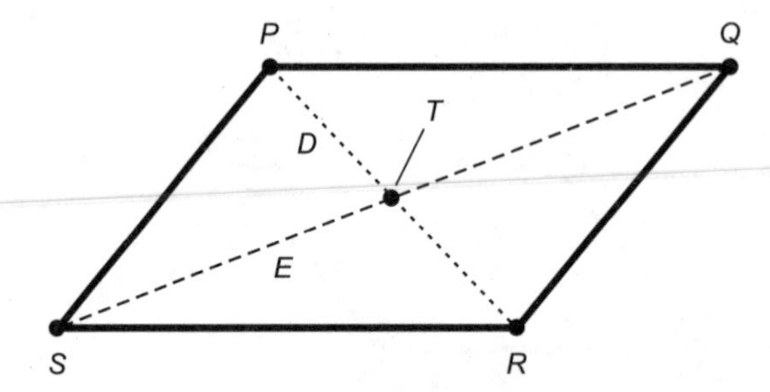

Fig. 3-9. The diagonals of a parallelogram bisect each other.

RECTANGLE

Suppose we have a parallelogram defined by four points P, Q, R, and S. Suppose any of the following statements is true for angles in degrees:

$$\angle QRS = 90° = \pi/2 \text{ rad}$$
$$\angle RSP = 90° = \pi/2 \text{ rad}$$
$$\angle SPQ = 90° = \pi/2 \text{ rad}$$
$$\angle PQR = 90° = \pi/2 \text{ rad}$$

Then all four interior angles are right angles, and the parallelogram is a *rectangle*: a four-sided plane polygon whose interior angles are all congruent. The converse of this is also true: If a quadrilateral is a rectangle, then any given interior angle is a right angle. Figure 3-10 shows an example of a parallelogram $PQRS$ in which $\angle QRS = 90° = \pi/2$ rad. Because one angle is a right angle and opposite pairs of sides are parallel, all four of the angles must be right angles.

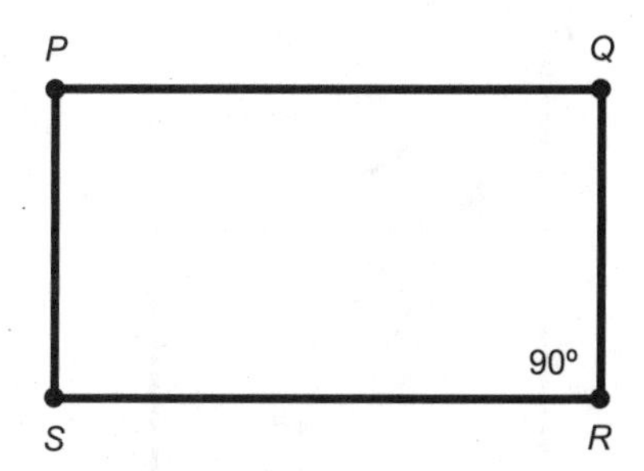

Fig. 3-10. If a parallelogram has one right interior angle, then the parallelogram is a rectangle.

RECTANGLE DIAGONALS

Suppose we have a parallelogram defined by four points P, Q, R, and S. Let D be the diagonal connecting P and R; let E be the diagonal connecting Q and S. Let the length of D be denoted by d; let the length of E be denoted by e (Fig. 3-11). If $d = e$, then the parallelogram is a rectangle. The converse is also true: if a parallelogram is a rectangle, then $d = e$. Thus, a parallelogram is a rectangle if and only if its diagonals have equal lengths.

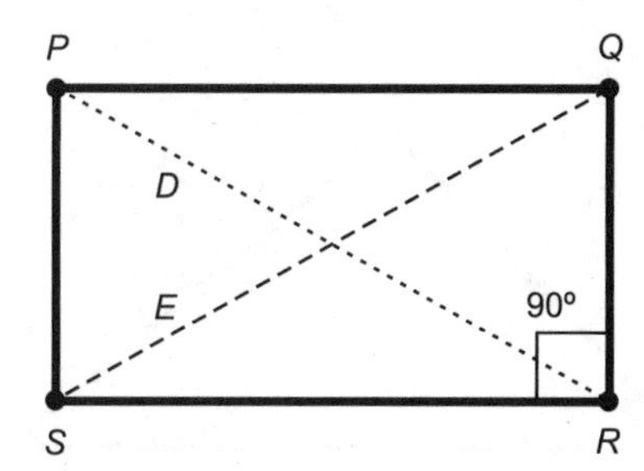

Fig. 3-11. The diagonals of a rectangle have equal length.

RHOMBUS DIAGONALS

Suppose we have a parallelogram defined by four points P, Q, R, and S. Let D be the diagonal connecting P and R; let E be the diagonal connecting Q and S. If D is perpendicular to E, then the parallelogram is a rhombus (Fig. 3-12). The converse is also true: If a parallelogram is a rhombus, then D is perpendicular to E. A parallelogram is a rhombus if and only if its diagonals are perpendicular.

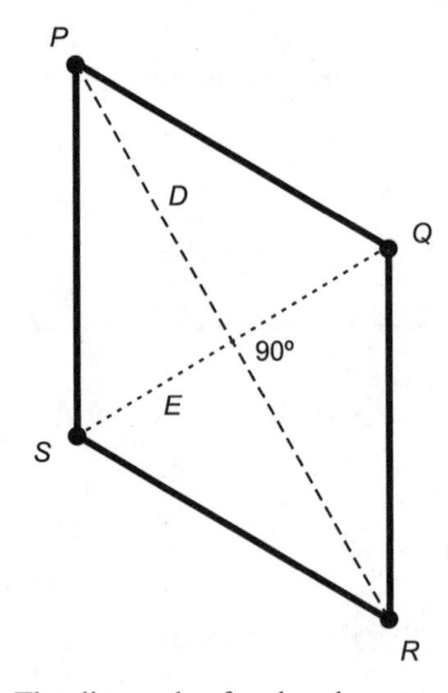

Fig. 3-12. The diagonals of a rhombus are perpendicular.

TRAPEZOID WITHIN TRIANGLE

Suppose we have a triangle defined by three points P, Q, and R. Let S be the midpoint of side PR, and let T be the midpoint of side PQ. Then line segments ST and RQ are parallel, and the figure defined by $STQR$ is a trapezoid

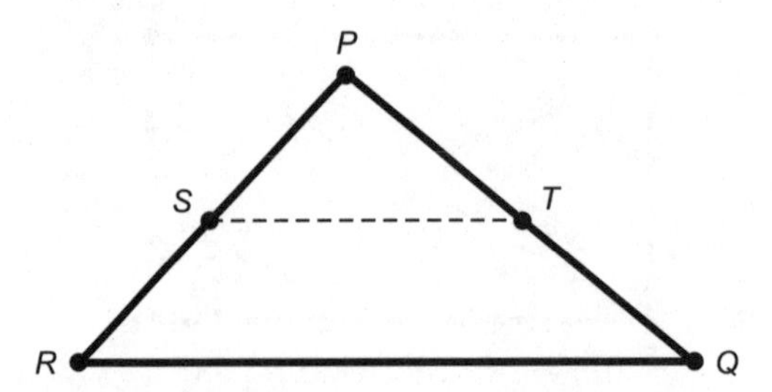

Fig. 3-13. A trapezoid is formed by "chopping off" the top of a triangle.

(Fig. 3-13). In addition, the length of line segment *ST* is half the length of line segment *RQ*.

MEDIAN OF A TRAPEZOID

Suppose we have a trapezoid defined by four points *P*, *Q*, *R*, and *S*. Let *T* be the midpoint of side *PS*, and let *U* be the midpoint of side *QR*. Line segment *TU* is called the *median* of trapezoid *PQRS*. The median of a trapezoid is always parallel to both the base and the top, and always splits the trapezoid into two other trapezoids. That is, polygons *PQUT* and *TURS* are both trapezoids (Fig. 3-14). In addition, the length of line segment *TU* is half the sum of the lengths of line segments *PQ* and *SR*. That is, the length of *TU* is equal to the average, or *arithmetic mean*, of the lengths of *PQ* and *SR*.

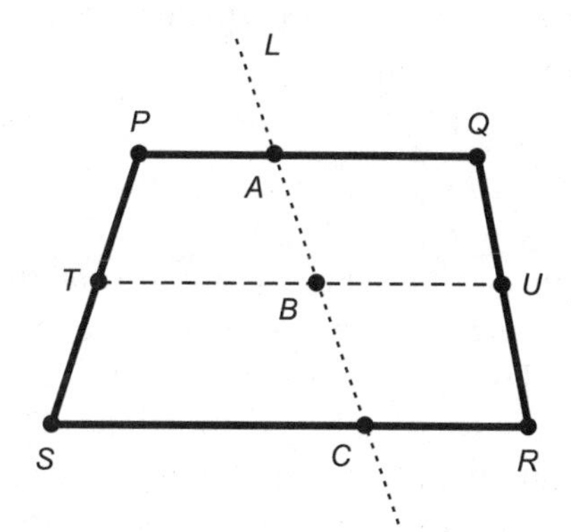

Fig. 3-14. The median of a trapezoid, also showing a transversal line.

MEDIAN WITH TRANSVERSAL

Look again at Fig. 3-14. Suppose *L* is a transversal line that crosses both the top of the large trapezoid (line segment *PQ*) and the bottom (line segment *SR*). Then *L* also crosses the median, line segment *TU*. Let *A* be the point at which *L* crosses *PQ*, let *B* be the point at which *L* crosses *TU*, and let *C* be the point at which *L* crosses *SR*. Then the lengths of line segments *AB* and *BC* are equal.

There is a second fact that should also be mentioned. Again, refer to Fig. 3-14. Suppose *PQRS* is a trapezoid, with sides *PQ* and *RS* parallel. Suppose *TU* is a line segment parallel to both *PQ* and *RS*, and that intersects both of the non-parallel sides of the trapezoid, that is, sides *PS* and *QR*. Let *L* be a transversal line that crosses all three parallel line segments *PQ*, *TU*, and *RS*, at the points A, B, and C respectively, as shown. In this scenario, if line segments AB and BC are equally long, then line segment TU is the median of the large trapezoid $PQRS$.

PROBLEM 3-3
Suppose a particular plane figure has diagonals that are the same length, and in addition, they intersect at right angles. What can be said about this polygon?

SOLUTION 3-3
From the above rules, this polygon must be a rectangle, because its diagonals are the same length. But it must also be a rhombus, because its diagonals are perpendicular to each other. There's only one type of polygon that can be both a rectangle and a rhombus, and that is a square. A square is a rhombus in which both pairs of opposite interior angles happen to have the same measure. A square is also a rectangle in which both pairs of opposite sides happen to be equally long.

PROBLEM 3-4
Suppose a sign manufacturing company gets tired of making rectangular billboards, and decides to put up a trapezoidal billboard instead. The top and the bottom of the billboard are horizontal, but neither of the other sides is vertical. The big sign measures 20 meters across the top edge, and 30 meters across the bottom edge. Two different companies want to advertise on the billboard, and both of them insist on having portions of equal height. What is the length of the line that divides the spaces allotted to the two advertisements? Does this represent a fair division of the sign area?

SOLUTION 3-4
The line segment that divides the two portions is the median of the sign. Its length, therefore, is the average of 20 meters and 30 meters, which, as you should be able to guess right away, is 25 meters. Whether or not this represents a fair split of the sign area can be debated. The advertiser on the bottom gets more area than the advertiser on the top, but the ad on top is likely to be the one that drivers in passing cars and trucks look at first. By the time drivers are finished with the ad on the top, they might be passing the sign.

Perimeters and Areas

Interior area is an expression of the size of the region enclosed by a polygon, and that lies in the same plane as all the vertices of the polygon. It is expressed in square units (or units squared). The *perimeter* of a polygon is the sum of the lengths of its sides. Perimeter can also be defined as the distance once around a polygon, starting at some point on one of its sides and proceeding clockwise or counterclockwise along the sides until that point is encountered again. Perimeter is expressed in linear units (or, if you prefer, "plain old units").

PERIMETER OF PARALLELOGRAM

Suppose we have a parallelogram defined by points P, Q, R, and S, with sides of lengths d and e as shown in Fig. 3-15. The two angles labeled x have equal measure. Let d be the base length and let h be the height. Then the perimeter, B, of the parallelogram is given by the following formula:

$$B = 2d + 2e$$

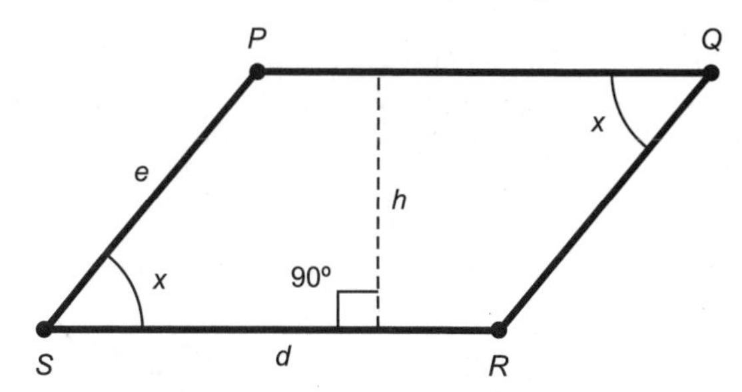

Fig. 3-15. Perimeter and area of parallelogram. A parallelogram is a rhombus if and only if $d = e$.

INTERIOR AREA OF PARALLELOGRAM

Suppose we have a parallelogram as defined above and in Fig. 3-15. The interior area, A, is the product of the base length and the height:

$$A = dh$$

PERIMETER OF RHOMBUS

Suppose we have a rhombus defined by points P, Q, R, and S, and having sides all of which have the same length. The rhombus is a special case of the parallelogram (Fig. 3-15) in which $d = e$. Let the lengths of all four sides be denoted d. The perimeter, B, of the rhombus is given by the following formula:

$$B = 4d$$

INTERIOR AREA OF RHOMBUS

Suppose we have a rhombus as defined above and in Fig. 3-15, where $d = e$. Let the lengths of all four sides be denoted d. The interior area, A, of the rhombus is the product of the length of any side and the height:

$$A = dh$$

PERIMETER OF RECTANGLE

Suppose we have a rectangle defined by points P, Q, R, and S, and having sides of lengths d and e as shown in Fig. 3-16. Let d be the base length, and let e be the height. The perimeter, B, of the rectangle is given by the following formula:

$$B = 2d + 2e$$

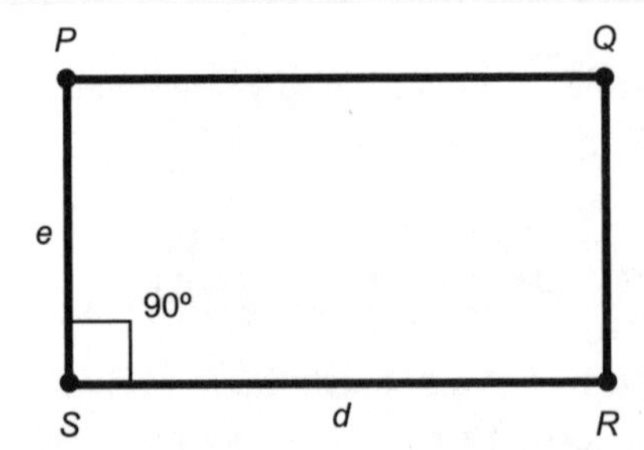

Fig. 3-16. Perimeter and area of rectangle. The figure is a square if and only if $d = e$.

INTERIOR AREA OF RECTANGLE

Suppose we have a rectangle as defined above and in Fig. 3-16. The interior area, A, is given by:

$$A = de$$

PERIMETER OF SQUARE

Suppose we have a square defined by points P, Q, R, and S, and having sides all of which have the same length. The square is a special case of the rectangle (Fig. 3-16) in which $d = e$. Let the lengths of all four sides be denoted d. The perimeter, B, of the square is given by the following formula:

$$B = 4d$$

INTERIOR AREA OF SQUARE

Suppose we have a square as defined above and in Fig. 3-16, where $d = e$. Let the lengths of all four sides be denoted d. The interior area, A, is equal to the square of the length of any side:

$$A = d^2$$

PERIMETER OF TRAPEZOID

Suppose we have a trapezoid defined by points P, Q, R, and S, and having sides of lengths d, e, f, and g as shown in Fig. 3-17. Let d be the base length, let h be the height, let x be the angle between the sides having length d and e, and let y be the angle between the sides having length d and g. Suppose the sides having lengths d and f (line segments RS and PQ) are parallel. Then the perimeter, B, of the trapezoid is:

$$B = d + e + f + g$$

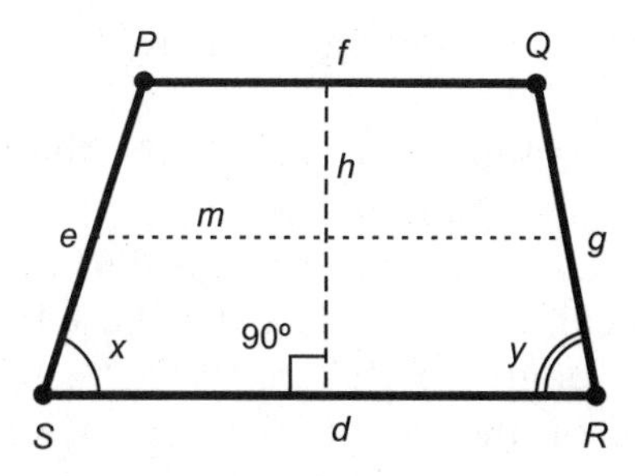

Fig. 3-17. Perimeter and area of trapezoid. Dimensions and angles are discussed in the text.

INTERIOR AREA OF TRAPEZOID

Suppose we have a trapezoid as defined above and in Fig. 3-17. The interior area, A, is equal to the average (or *arithmetic mean*) of the lengths of the base and the top, multiplied by the height. The formula for calculating A is as follows:

$$A = [(d + f)/2]h$$
$$= (dh + fh)/2$$

Suppose m represents the length of the median of the trapezoid, that is, a line segment parallel to the base and the top, and midway between them. Then the interior area is equal to the product of the length of the median and the height:

$$A = mh$$

PROBLEM 3-5

Refer back to Problem 3-4. Suppose the whole billboard is 15 meters high. Recall that it is a trapezoidal billboard, measuring 20 meters along the top edge and 30 meters along the bottom. The sign is divided by a median, horizontally placed midway between the top and the bottom. What fraction of the total billboard surface area, as a percentage, does the advertiser with the top half get?

SOLUTION 3-5

The length of the median, as determined in Problem 3-4, is 25 meters, which is the average of the lengths of the bottom and the top. Thus $m = 25$. We are given that $h = 15$. The total interior area of the sign, call it A_{total}, is therefore:

$$A_{total} = 25 \text{ meters} \times 15 \text{ meters}$$
$$= 375 \text{ meters squared}$$

The area of the top half is found by considering the trapezoid in which m forms the base. We must use the more complicated formula—the one involving the arithmetic mean, above—in order to find the interior area of this smaller trapezoid. Let's call this area A_{top}. The base length of this trapezoid is 25 meters, while the length of the top is 20 meters. The height is 7.5 meters, half the height of the whole sign. Thus, A_{top} is found by this calculation:

$$A_{top} = [(25 \text{ meters} + 20 \text{ meters})/2] \times 7.5 \text{ meters}$$
$$= (45 \text{ meters}/2) \times 7.5 \text{ meters}$$
$$= 22.5 \text{ meters} \times 7.5 \text{ meters}$$
$$= 168.75 \text{ meters squared}$$

The fraction of the total area represented by the top portion of the sign is the ratio of A_{top} to A_{total}. That is 168.75 meters squared divided by 375 meters squared, or 0.45. Therefore, the top advertiser gets 45 percent of the total interior area of the sign.

PROBLEM 3-6

Suppose the billboard is a rectangle rather than a trapezoid, measuring 25 meters across both the top and the bottom. Suppose the sign is 15 meters tall, and is to be split into upper and lower portions, one for each of two different advertisers, Top Inc. and Bottom Inc. Suppose that the executives of Bottom Inc. demand that Top Inc. only get 45 percent of the total area of the sign because of Top Inc.'s more favorable viewing position. How far from the bottom of the sign should the dividing line be placed?

SOLUTION 3-6

The total area of the sign, A_{total}, is equal to the product of the base (or top) length and the height:

$$A_{total} = 25 \text{ meters} \times 15 \text{ meters} = 375 \text{ meters squared}$$

This is the same total area as that found in Solution 3-5. Thus, 45 percent of this, A_{top}, is the same as in Solution 3-5, that is, 168.75 meters squared. This means that the area of the bottom portion, A_{bottom}, is:

$$A_{bottom} = A_{total} - A_{top}$$
$$= (375 - 168.75) \text{ meters squared}$$
$$= 206.25 \text{ meters squared}$$

Let x be the distance, in meters, that the dividing line is to be placed from the bottom edge of the sign. Then x represents the lengths of the two vertical sides of the bottom rectangle. We already know that the dividing line (which is the top edge of the bottom rectangle) is 25 meters long, as is the base. So we get this formula:

$$A_{bottom} = 25x$$

We know that $A_{bottom} = 206.25$ meters squared. So we can plug this into the above equation and solve for x:

$$206.25 = 25x$$
$$x = (206.25 \text{ meters squared})/(25 \text{ meters})$$
$$= 8.25 \text{ meters}$$

The dividing line should therefore be placed 8.25 meters above the bottom edge of the billboard. Is this placement fair? That will have to be determined by mutual discussions between the lawyers for Top Inc. and Bottom Inc., doubtless at shareholder expense.

Quiz

Refer to the text in this chapter if necessary. A good score is eight correct. Answers are in the back of the book.

1. A quadrilateral cannot have
   (a) four sides of unequal length
   (b) four interior angles of unequal measure
   (c) any sides with zero length
   (d) an interior angle whose measure is more than 180°

2. A rhombus has sides that each measure 4 units in length. The interior area of this quadrilateral is
   (a) 4 units
   (b) 8 units
   (c) 16 units
   (d) impossible to determine without more information

3. The median of a trapezoid
   (a) is parallel to the shortest side
   (b) is perpendicular to the longest side
   (c) is parallel to the base
   (d) is perpendicular to the base

4. Suppose we are told two things about a quadrilateral: first, that it is a parallelogram, and second, that one of its interior angles measures 60°. The measure of the angle adjacent to the 60° angle is
   (a) 60°
   (b) 90°
   (c) 120°
   (d) impossible to know without more information

5. In the scenario of Question 4, the measure of the angle opposite the 60° angle is
   (a) 60°
   (b) 90°
   (c) 120°
   (d) impossible to know without more information

6. One of the interior angles of a quadrilateral measures $3\pi/2$ rad. From this alone, we know that this quadrilateral
   (a) cannot lie in a single plane
   (b) must be a parallelogram
   (c) must be a trapezoid
   (d) cannot be a parallelogram or a trapezoid

7. A parallelogram is a special instance of
   (a) a rectangle
   (b) a trapezoid
   (c) a rhombus
   (d) a square

8. A square is a special instance of
   (a) a parallelogram
   (b) a rectangle
   (c) a rhombus
   (d) more than one of the above

9. The length of the median of a trapezoid is
   (a) equal to the average of the lengths of the two sides parallel to it
   (b) equal to the difference between the lengths of the two sides parallel to it
   (c) equal to the average of the lengths of the two sides it intersects
   (d) equal to the difference between the lengths of the two sides it intersects

10. Suppose a square has a diagonal measure of 10 units. The area of the square is
    (a) 25 square units
    (b) 50 square units
    (c) 100 square units
    (d) impossible to determine without more information

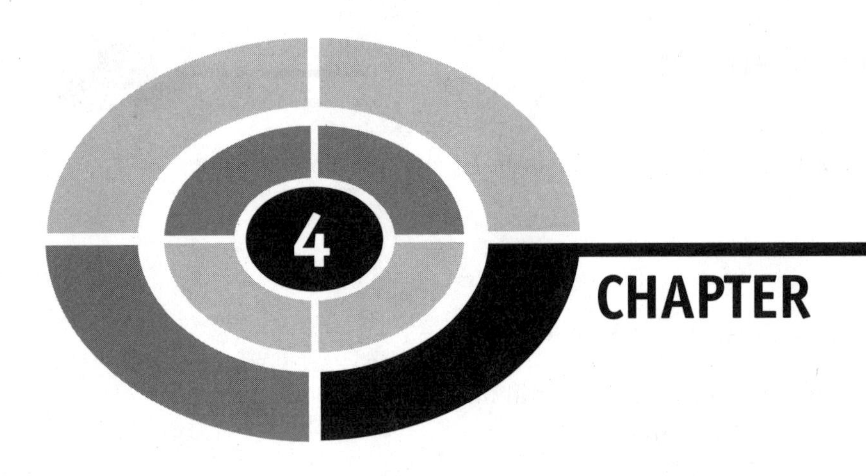

Other Plane Figures

There is no limit to the number of sides a polygon can have. In order to qualify as a plane polygon, all of the vertices (points where the sides come together) must lie in the same plane, and no two sides are allowed to cross over each other. No two vertices can coincide. No three vertices can lie on a common line (otherwise we might get confused as to whether a line segment represents one side or two). And finally, the sides must all be straight line segments having finite length. They can't be curved, and they can't go off into infinity.

Five Sides and Up

As you can guess, plane polygons get increasingly complicated as the number of sides increases. Let's consider a few special cases.

THE REGULAR PENTAGON

Figure 4-1 shows a five-sided polygon, all of whose sides have the same length, and all of whose interior angles have the same measure. This is called

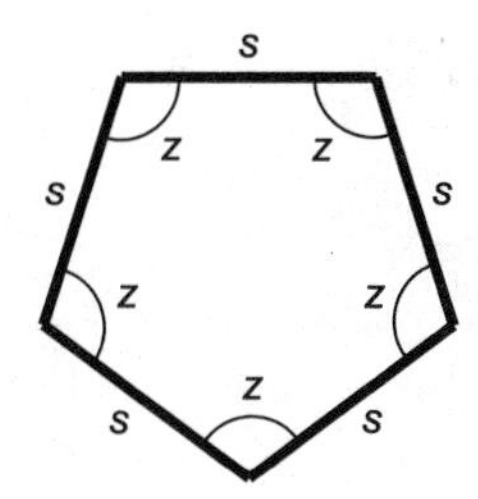

Fig. 4-1. A regular pentagon. Each side is s units long, and each interior angle has measure z.

a *regular pentagon*. It is called *convex* because its exterior never bends inward. Another way of saying this is that all of the interior angles measure less than 180° (π rad).

THE REGULAR HEXAGON

A convex polygon with six sides, all of which are equally long, is called a *regular hexagon* (Fig. 4-2). This type of polygon is common in nature. If there are many of them and they are all the same size, they can be placed neatly together without any gaps. (Do you remember those old barbershops where the floors were made of little hexagonal tiles that fit up against each other?) This makes the regular hexagon a special sort of figure, along with the equilateral triangle and the square. Certain crystalline solids form regular hexagonal shapes when they fracture. Snowflakes have components with this shape.

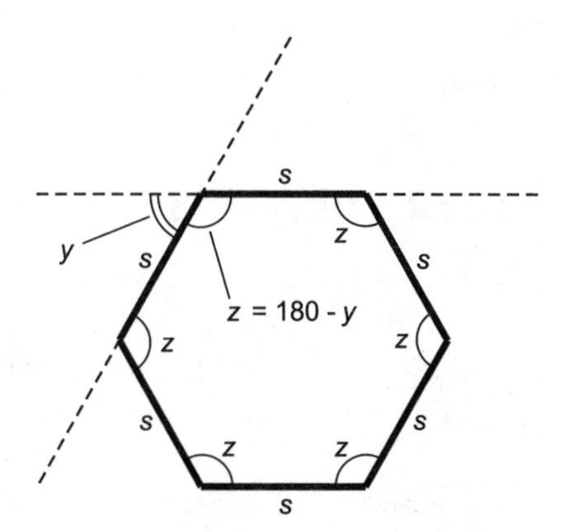

Fig. 4-2. A regular hexagon. Each side is s units long, and each interior angle has measure z. The extensions of sides (dashed lines) are the subject of Problem 4-1.

THE REGULAR OCTAGON

Figure 4-3 shows a *regular octagon*. This is a convex polygon with eight sides, all equally long. As is the case with the regular hexagon, large numbers of these figures can be fit neatly together. So it is not surprising that nature has seen fit to take advantage of this, building octagonal crystals.

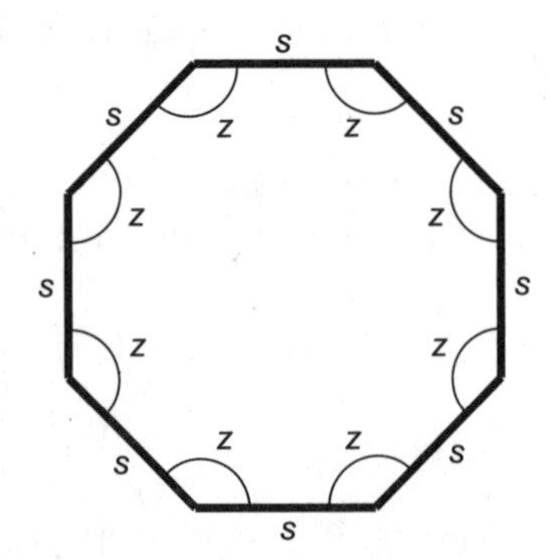

Fig. 4-3. A regular octagon. Each side is s units long, and each interior angle has measure z.

REGULAR POLYGONS IN GENERAL

For every whole number n greater than or equal to 3, it is possible to have a regular polygon with n sides. So far we've seen the equilateral triangle ($n = 3$), the square ($n = 4$), the regular pentagon ($n = 5$), the regular hexagon ($n = 6$), and the regular octagon ($n = 8$). There can exist a regular polygon with 1000 sides (this might be called a "regular kilogon"), 1,000,000 sides (a "regular megagon"), or 1,000,000,000 sides (a "regular gigagon"). These last three would look pretty much like circles to the casual observer.

GENERAL, MANY-SIDED POLYGONS

Once the restrictions are removed concerning the relationship among the sides of a polygon having four sides or more, the potential for variety increases without limit. Sides can have all different lengths, and the measure of each interior angle can range anywhere from 0° (0 rad) to 360° (2π rad), non-inclusive.

Figure 4-4 shows some examples of general, many-sided polygons. The object at the top left is a non-convex octagon whose sides happen to all have the same length. The interior angles, however, differ in measure. The other

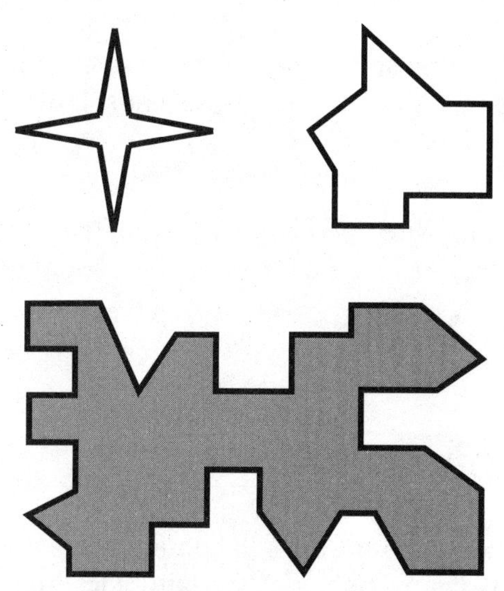

Fig. 4-4. General, many-sided polygons. The object with the shaded interior is the subject of Problem 4-2.

two objects are irregular and non-convex. All three share the essential characteristics of a plane polygon:

- The vertices all lie in a single plane
- No two sides cross
- No two vertices coincide
- No three vertices lie on a single straight line
- All the sides are line segments of finite length

PROBLEM 4-1

What is the measure of each interior angle of a regular hexagon?

SOLUTION 4-1

Draw a horizontal line segment to start. All the other sides must be duplicates of this one, but rotated with respect to the first line segment by whole-number multiples of a certain angle. This rotation angle from side to side is 360° divided by 6 (a full rotation divided by the number of sides), or 60°. Imagine the lines on which two adjacent sides lie. Look back at Fig. 4-2. These lines subtend a 60° angle with respect to each other, if you look at the acute angle. But if you look at the obtuse angle, it is 120°. This obtuse angle is an interior angle of the hexagon. Therefore, each interior angle of a regular hexagon measures 120°.

PROBLEM 4-2

Briefly glance at the lowermost polygon in Fig. 4-4 (the one with the shaded interior). Don't look at it for more than two seconds. How many sides do you suppose this object has?

SOLUTION 4-2

This is an optical illusion. Most people underestimate the number of sides in figures like this. After you've made your guess, count them and see for yourself!

Some Rules of "Polygony"

All plane polygons share certain things in common. It's possible to calculate the perimeter or area of any polygon. Certain rules and definitions apply concerning the interior and exterior angles, and the relationships between the angles and the sides. Some of the more significant rules of "polygony" (pronounced "pa-LIG-ah-nee"), a make-believe term that means "the science of polygons," follow.

IT'S GREEK TO US

Mathematicians, scientists, and engineers often use Greek letters to represent geometric angles. The most common symbol for this purpose is an italicized, lowercase Greek letter theta (pronounced "THAY-tuh"). It looks like a numeral zero leaning to the right, with a horizontal line across its middle (θ).

When writing about two different angles, a second Greek letter is used along with θ. Most often, it is the italicized, lowercase letter phi (pronounced "fie" or "fee"). It looks like a lowercase English letter o leaning to the right, with a forward slash through it (ϕ). You might as well get used to these symbols, because if you have anything to do with engineering and science, you're going to encounter them.

Sometimes the italic, lowercase Greek alpha ("AL-fuh"), beta ("BAY-tuh"), and gamma ("GAM-uh") are used to represent angles. These, respectively, look like this: α, β, γ. When things get messy and there are a lot of angles to talk about, numeric subscripts are sometimes used, so don't be surprised if you see angles denoted θ_1, θ_2, θ_3, and so on.

SUM OF INTERIOR ANGLES

Let V be a plane polygon having n sides. Let the interior angles be θ_1, θ_2, θ_3, ..., θ_n (Fig. 4-5). The following equation holds if the angular measures are given in degrees:

$$\theta_1 + \theta_2 + \theta_3 + \cdots + \theta_n = 180n - 360 = 180(n - 2)$$

If the angular measures are given in radians, then the following holds:

$$\theta_1 + \theta_2 + \theta_3 + \cdots + \theta_n = \pi n - 2\pi = \pi(n - 2)$$

In these examples, the degree symbol (°) and the radian abbreviation (rad) are left out for simplicity. It is all right to do this, as long as it is clear which angular units we're dealing with.

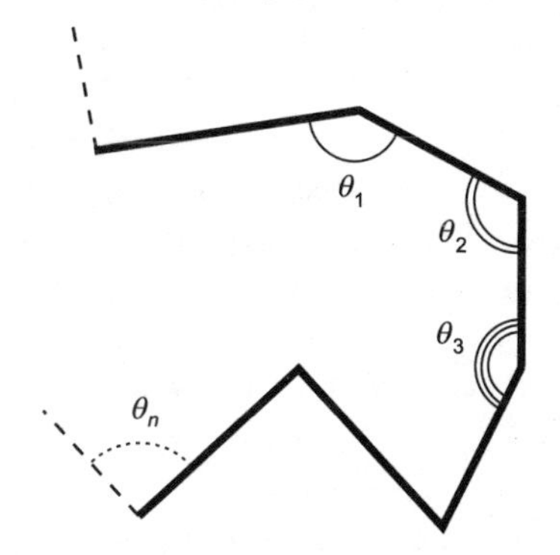

Fig. 4-5. Adding up the measures of the interior angles of a general, many-sided polygon.

INDIVIDUAL INTERIOR ANGLES OF REGULAR POLYGON

Let V be a plane polygon having n sides whose interior angles all have equal measure given by θ, and whose sides all have equal length given by s (Fig. 4-6). Then V is a regular polygon, and the measure of each interior angle, θ, in degrees is given by the following formula:

$$\theta = (180n - 360)/n$$

If the angular measures are given in radians, then the formula looks like this:

$$\theta = (\pi n - 2\pi)/n$$

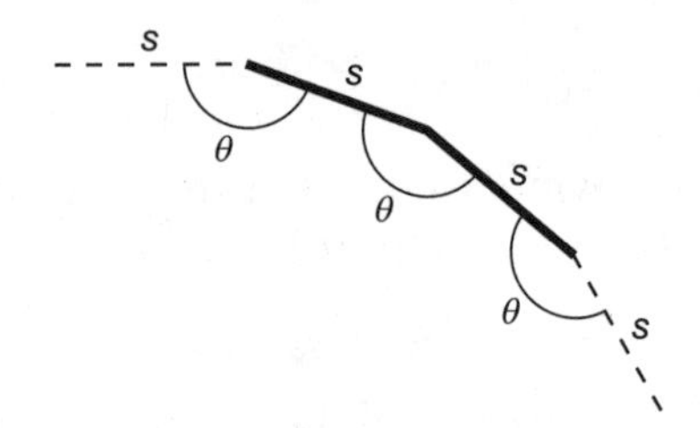

Fig. 4-6. Interior angles of a regular, many-sided polygon.

POSITIVE AND NEGATIVE EXTERIOR ANGLES

An *exterior angle* of a polygon is measured counterclockwise between a specific side and the extension of a side next to it. An example is shown in Fig. 4-7. If the arc of the angle lies outside the polygon, then the resulting angle θ has a measure between, but not including, 0 and 180 degrees. The angle is positive because it is measured "positively counterclockwise":

$$0^\circ < \theta < 180^\circ$$

If the arc of the angle lies inside the polygon, then the angle is measured clockwise ("negatively counterclockwise"). This results in an angle ϕ with a measure between, but not including, –180 and 0 degrees:

$$-180^\circ < \phi < 0^\circ$$

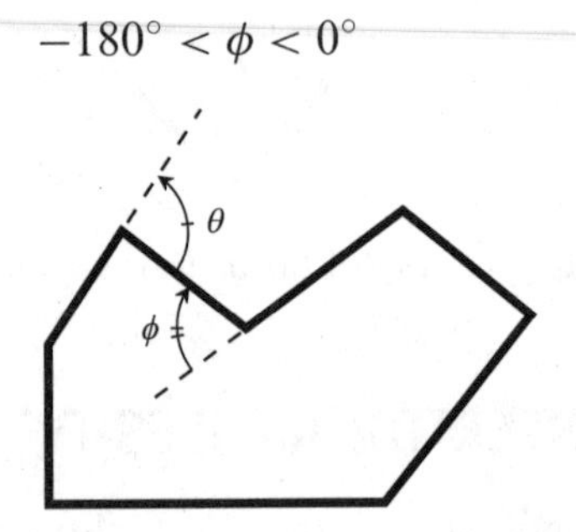

Fig. 4-7. Exterior angles of an irregular polygon. The angle θ is measured "positively counterclockwise" while the angle ϕ is measured "negatively counterclockwise."

PERIMETER OF REGULAR POLYGON

Let V be a regular plane polygon having n sides of length s, and whose vertices are P_1, P_2, P_3, ..., P_n as shown in Fig. 4-8. Then the perimeter, B, of the polygon is given by the following formula:

$$B = ns$$

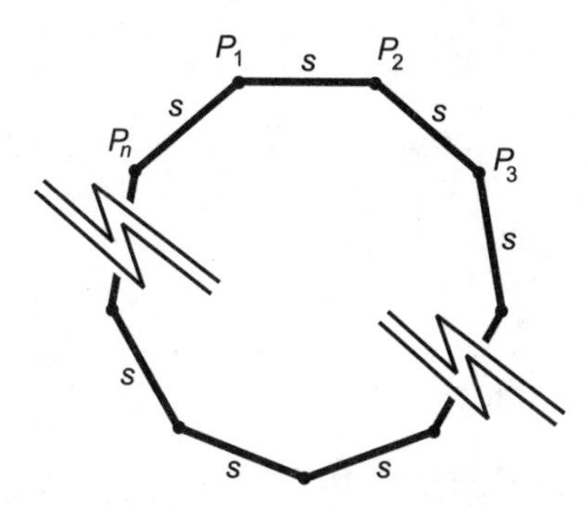

Fig. 4-8. Perimeter and area of a regular, n-sided polygon. Vertices are labeled P_1, P_2, P_3, ..., P_n. The length of each side is s units.

A TASTE OF TRIGONOMETRY

Some of the following rules involve *trigonometry*. This branch of mathematics has an undeserved bad reputation among some students. According to various rumors, trigonometry is esoteric (this is not true), is inherently incomprehensible (also not true), was dreamed up by sadistic theoreticians with the intent of confusing people of ordinary intelligence (unlikely, but no one knows for sure), and is peppered with Greek symbology (well, yes). There are six *trigonometric functions*, also known as *circular functions*. They are the *sine*, *cosine*, *tangent*, *cosecant*, *secant*, and *cotangent*. All six of these functions produce specific numbers at their "outputs" when certain angular measures are fed into their "inputs."

We won't concern ourselves here with formal definitions of the circular functions, or how they are derived. All you need to know in order to use the following rules is how to use the sine and cosine function keys on a calculator. The sine of an angle is found by entering the angle's measure in degrees or radians into a calculator, and then hitting the "sine" or "sin" function key. The cosine is found by entering the angle's measure in degrees or radians and then hitting "cosine" or "cos." Some calculators have a "tangent" or "tan" function key, and others don't. If your calculator doesn't have a tangent key, the tangent of an angle can be found by dividing its sine by its cosine. Many calculators lack a "cotangent" or "cot" key, but the cotangent of an angle is equal to the reciprocal of its tangent, or the cosine divided by the sine.

INTERIOR AREA OF REGULAR POLYGON

Let V be a regular, n-sided polygon, each of whose sides have length s as defined above and in Fig. 4-8. The interior area, A, is given by the following formula if angles are specified in degrees:

$$A = (ns^2/4) \cot (180/n)$$

If angles are specified in radians, then:

$$A = (ns^2/4) \cot (\pi/n)$$

PROBLEM 4-3
What is the interior area of a regular, 10-sided polygon, each of whose sides is exactly 2 units long? Express your answer to two decimal places.

SOLUTION 4-3
In this case, $n = 10$ and $s = 2$. Let's use degrees for the angles. Then we can plug our values of n and s into the first formula, above, getting this:

$$\begin{aligned} A &= (10 \times 2^2/4) \cot (180/10) \\ &= (10 \times 4/4) \cot 18 \\ &= 10 \cot 18 \\ &= 10 \cos 18/\sin 18 \\ &= 10 \times 0.951057/0.309017 \\ &= 10 \times 3.07769 \\ &= 30.7769 \\ &= 30.78 \text{ square units (to two decimal places)} \end{aligned}$$

In order to obtain an answer to two decimal places, it's best to use five or six decimal places throughout the calculation, rounding off only at the end. This will ensure that cumulative errors are kept to a minimum.

PROBLEM 4-4
What is the interior area of a regular, 100-sided polygon, each of whose sides is exactly 0.20 units long? Express your answer to two decimal places.

SOLUTION 4-4
In this example, $n = 100$ and $s = 0.20$. If you're astute, you'll notice that the perimeter of this polygon is $100 \times 0.20 = 20$ units, the same as the perimeter of the 10-sided polygon of Problem 4-3, which is $10 \times 2.0 = 20$ units. Imagine these two regular polygons sitting side-by-side. Draw approximations of them if you like. It seems reasonable to suppose that the area of the 100-sided

polygon should be a little larger than that of the 10-sided figure, but not much larger. Let's find out. For fun, let's use radians instead of degrees this time.

Be sure your calculator is set to work with radians, not degrees, before each and every use of a trigonometric function key. Here we go:

$$\begin{aligned}
A &= (100 \times 0.20^2/4) \cot (\pi/100) \\
&= (100 \times 0.04/4) \cot 0.0314159 \\
&= 1 \times \cot 0.0314159 \\
&= \cot 0.0314159 \\
&= \cos 0.0314159/\sin 0.0314159 \\
&= 0.999507/0.031411 \\
&= 31.8203 \\
&= 31.82 \text{ square units (to two decimal places)}
\end{aligned}$$

When doing calculations like the ones above, it's important to go through each step twice. Alternatively, you can go through the entire process twice. Best of all, take both of these precautions! It is amazing how many errors can be made by humans using calculators to crunch numbers. The most common mistakes occur as a result of a failure to hit the function keys in the right order.

Circles and Ellipses

A *circle* is a geometric figure consisting of all points in a plane that are equally distant from some center point. Imagine a flashlight with a round lens that throws a brilliant central beam of light surrounded by a dimmer cone of light. Suppose you switch this flashlight on, and point it straight down at the floor in a dark room. The outline of the dim light cone is a circle. If you turn the flashlight so the entire dim light cone lands on the floor but the brilliant central light ray is not pointed straight down, the outline of the dim light cone is an *ellipse*.

The circle and the ellipse are examples of *conic sections*. This term arises from the fact that both the circle and the ellipse can be defined as sets of points resulting from the intersection of a plane with a cone.

A SPECIAL NUMBER

The circumference of a circle, divided by its diameter in the same units, is a constant that does not depend on the size of a circle. This fact was noticed by mathematicians thousands of years ago. The value of this number cannot be expressed as a ratio of whole numbers. For this reason, this number is called an *irrational number*. ("Irrational" means, in this context, "having no ratio.") If you try to write this number in decimal form, you get a non-terminating, non-repeating sequence of digits after the decimal point. It is a constant called pi, and is symbolized π. This is the same π we encountered earlier when defining the radian as a unit of angular measure.

The value of π has been calculated to many millions of decimal places by supercomputers. It's approximately equal to 3.14159. If you need more accuracy, you can use the calculator function in a personal computer. In a computer that uses the Windows(TM) operating system, open the calculator program and set it for scientific mode. Check the box marked "Inv." Be sure there are black dots in both the "Dec" and "Radians" spaces. Press 1, then the minus key, then 2, and then the equals key so you get –1 on the display. Finally, hit the "cos" button. This will show you the angle, in radians, whose cosine is equal to –1; this happens to be π. A good calculator will display enough digits to make almost anyone happy.

Here are some formulas that can be used to find the perimeters and areas of circles, ellipses, and regular polygons that are inscribed within, or circumscribed around, circles. You don't have to memorize these (except for the formulas for the perimeter and interior area of a circle, which are worth memorizing), but they can be useful for reference. As with all the formulas in this book, they are straightforward, even if some of them look messy. Using them is simply a matter of entering numbers into a calculator and making sure you hit the correct keys in the correct order.

PERIMETER OF CIRCLE

Let C be a circle having radius r as shown in Fig. 4-9. Then the perimeter (or circumference), B, of the circle is given by the following formula:

$$B = 2\pi r$$

INTERIOR AREA OF CIRCLE

Let C be a circle as defined above and in Fig. 4-9. The interior area, A, of the circle is given by:

$$A = \pi r^2$$

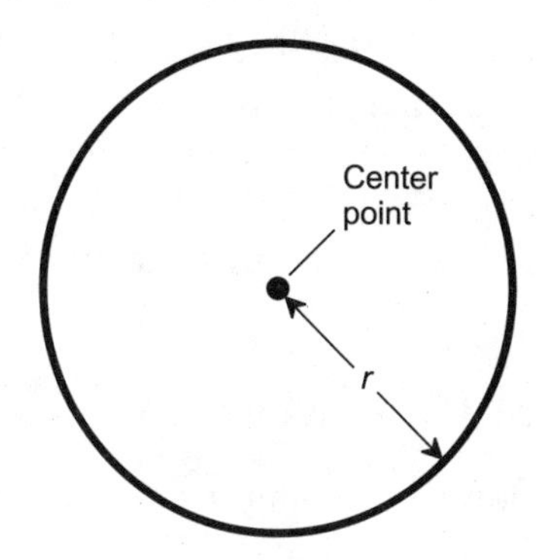

Fig. 4-9. Dimensions of a circle. The radius measures r units.

INTERIOR AREA OF ELLIPSE

Let E be an ellipse whose major (longer) semi-axis measures r_1 units and whose minor (shorter) semi-axis measures r_2 units, as shown in Fig. 4-10. The interior area, A, of the ellipse is given by:

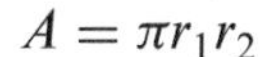

$$A = \pi r_1 r_2$$

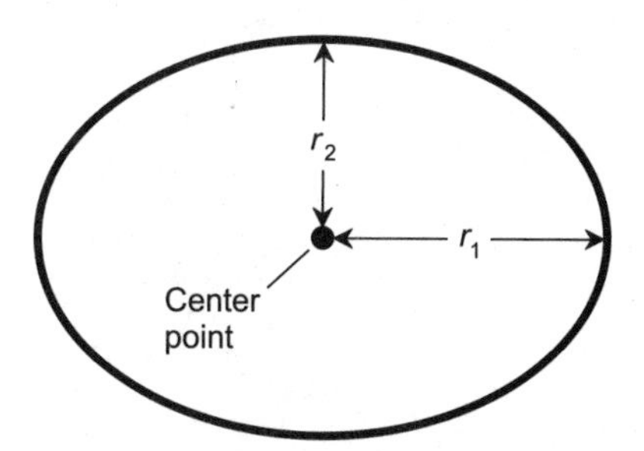

Fig. 4-10. Dimensions of an ellipse. The major semi-axis measures r_1 units, and the minor semi-axis measures r_2 units.

ELLIPTICITY

The ratio of the length of the major semi-axis to the length of the minor semi-axis is a quantitative indicator of the extent to which an ellipse is elongated. The number r_1/r_2 is called the *ellipticity*, and can be symbolized by the lowercase, italic Greek letter epsilon (ε). Thus:

$$\varepsilon = r_1/r_2$$

When $\varepsilon = 1$, an ellipse is a circle. Because r_1 is defined as the major (longer) semi-axis, ε is always greater than or equal to 1. Ellipticity should not be confused with *eccentricity*, another measure of the extent to which a curve deviates from a circle. Eccentricity is defined differently than ellipticity, and involves not only the circle and the ellipse, but other conic sections as well.

PERIMETER OF INSCRIBED REGULAR POLYGON

Let V be a regular plane polygon having n sides, and whose vertices P_1, P_2, P_3, ..., P_n lie on a circle of radius r (Fig. 4-11). Then the perimeter, B, of the polygon is given by the following formula when angles are specified in degrees:

$$B = 2nr \sin (180/n)$$

If angles are specified in radians, then:

$$B = 2nr \sin (\pi/n)$$

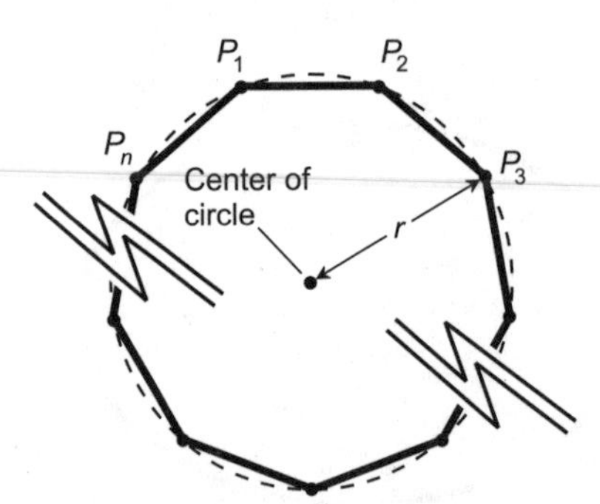

Fig. 4-11. Perimeter and area of inscribed regular polygon. The radius of the circle measures r units. Vertices of the polygon, all of which lie on the circle, are labeled P_1, P_2, P_3, ..., P_n.

INTERIOR AREA OF INSCRIBED REGULAR POLYGON

Let V be a regular polygon as defined above and in Fig. 4-11. The interior area, A, of the polygon is given by the following formula if angles are specified in degrees:

$$A = (nr^2/2) \sin (360/n)$$

If angles are specified in radians, then:

$$A = (nr^2/2) \sin (2\pi/n)$$

PERIMETER OF CIRCUMSCRIBED REGULAR POLYGON

Let V be a regular plane polygon having n sides whose center points P_1, P_2, P_3, ..., P_n lie on a circle of radius r (Fig. 4-12). The perimeter, B, of the polygon is given by the following formula when angles are specified in degrees:

$$B = 2nr \tan (180/n)$$

If angles are specified in radians, then:

$$B = 2nr \tan (\pi/n)$$

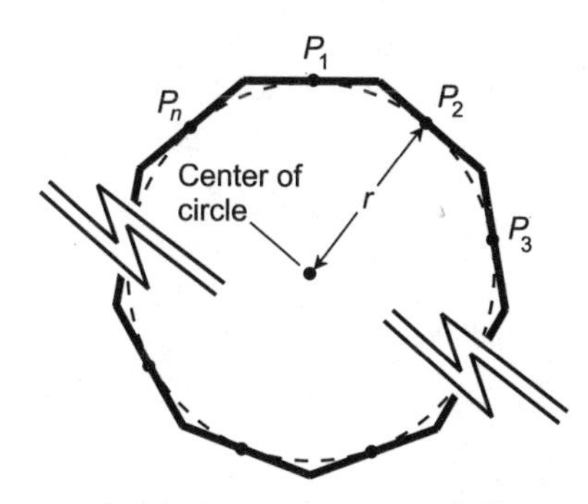

Fig. 4-12. Perimeter and area of circumscribed regular polygon. The radius of the circle measures r units. Center points of the sides of the polygon, all of which lie on the circle, are labeled P_1, P_2, P_3, ..., P_n.

INTERIOR AREA OF CIRCUMSCRIBED REGULAR POLYGON

Let V be a regular polygon as defined above and in Fig. 4-12. The interior area, A, of the polygon is given by the following formula if angles are specified in degrees:

$$A = nr^2 \tan (180/n)$$

If angles are specified in radians, then:

$$A = nr^2 \tan (\pi/n)$$

PERIMETER OF CIRCULAR SECTOR

Let S be a sector of a circle whose radius is r (Fig. 4-13). Let θ be the apex angle in radians. The perimeter, B, of the sector is given by the following formula:

$$B = r(2 + \theta) = 2r + r\theta$$

If θ is specified in degrees, then the perimeter, B, of the sector is given by:

$$B = 2r(1 + 90\theta)/\pi$$

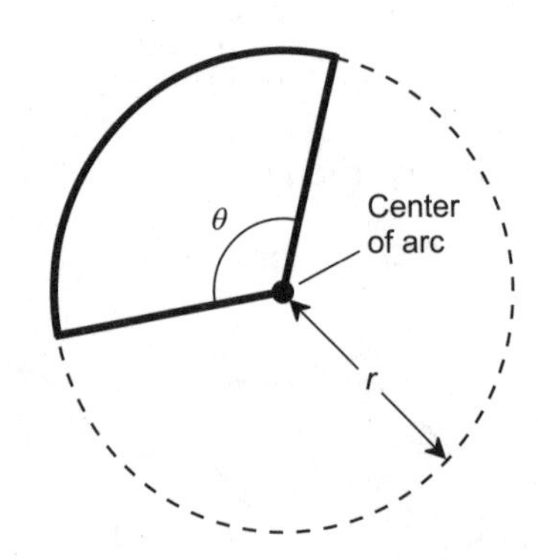

Fig. 4-13. Perimeter and area of circular sector. The radius of the circle measures r units, and the arc subtends an angle θ.

INTERIOR AREA OF CIRCULAR SECTOR

Let S be a sector of a circle as defined above and in Fig. 4-13. Let θ be the apex angle in radians. The interior area, A, of the sector is given by:

$$A = r^2\theta/2$$

If θ is specified in degrees, then the interior area, A, of the sector is given by:

$$A = 90r^2\theta/\pi$$

PROBLEM 4-5

What is the area of a regular octagon inscribed within a circle whose radius is exactly 10 units?

SOLUTION 4-5

Let's use the formula for the area of an inscribed regular polygon, where angles are expressed in degrees:

$$A = (nr^2/2) \sin (360/n)$$

where A is the area in square units, n is the number of sides in the polygon, and r is the radius of the circle. We know that $n = 8$ and $r = 10$, so we can plug in the numbers and use a calculator as needed:

$$\begin{aligned} A &= (8 \times 10^2/2) \sin (360/8) \\ &= 400 \sin 45^\circ \\ &= 400 \times 0.7071 \\ &= 283 \text{ square units (approximately)} \end{aligned}$$

PROBLEM 4-6

What is the perimeter of a regular 12-sided polygon circumscribed around a circle whose radius is exactly 4 units?

SOLUTION 4-6

Let's use the formula for the perimeter of a circumscribed regular polygon, where angles are expressed in radians:

$$B = 2nr \tan (\pi/n)$$

where B is the perimeter, n is the number of sides in the polygon, and r is the radius of the circle. Consider $\pi = 3.14159$. We know that $n = 12$ and $r = 4$. We plug in the numbers and use a calculator, being sure the angle function is set for radians, not for degrees:

$$\begin{aligned} B &= 2 \times 12 \times 4 \tan (\pi/12) \\ &= 96 \tan 0.261799 \\ &= 96 \times 0.26795 \\ &= 25.72 \text{ units (approximately)} \end{aligned}$$

PROBLEM 4-7

How would you expect the perimeter of the circumscribed polygon in Problem 4-6 to compare with the perimeter of the circle around which it is circumscribed?

SOLUTION 4-7

It is reasonable to suppose that the perimeter of the polygon is slightly greater than the perimeter (or circumference) of the circle. Let's calculate the circumference of the circle to see if this is true, and if so, to what extent. We use the formula for the circumference of a circle:

$$B = 2\pi r$$

where B is the circumference and r is the radius. We know $r = 4$, and we can consider $\pi = 3.14159$. Thus:

$$\begin{aligned} B &= 2 \times 3.14159 \times 4 \\ &= 25.13 \text{ units (approximately)} \end{aligned}$$

Quiz

Refer to the text in this chapter if necessary. A good score is eight correct. Answers are in the back of the book.

1. As the number of sides in a regular polygon increases without limit, assuming its interior area remains constant, the length of each side
   (a) increases without limit
   (b) approaches zero
   (c) stays the same
   (d) none of the above
2. As the number of sides in a regular polygon increases without limit, the measure of each interior angle
   (a) increases without limit
   (b) approaches 0°
   (c) stays the same
   (d) none of the above
3. Suppose X is the area of an n-sided, regular polygon circumscribed around a circle, and Y is the area of an n-sided, regular polygon inscribed within that same circle. As n becomes larger and larger without limit (or, as is sometimes said, n approaches infinity), what happens to the ratio of X to Y?
   (a) It approaches 1
   (b) It does not change
   (c) It grows without limit
   (d) We cannot say unless we know the area of the circle
4. An irregular, non-convex polygon can have some interior angles
   (a) that measure more than 180°

(b) that measure more than 360°
(c) that measure exactly 0°
(d) that measure exactly 360°

5. In an ellipse that is not a circle
   (a) the major and minor axes have equal length
   (b) the major axis is longer than the minor axis
   (c) the minor axis is longer than the major axis
   (d) none of the above statements is true

6. As the number of sides in a regular polygon increases without limit, the sum of the measures of its interior angles
   (a) increases without limit
   (b) decreases and approaches 0°
   (c) increases and approaches 360°
   (d) increases and approaches 720°

7. Each interior angle of a regular pentagon has a measure of
   (a) $2\pi/5$ rad
   (b) $3\pi/5$ rad
   (c) $5\pi/3$ rad
   (d) $5\pi/2$ rad

8. A circle has a radius of 2 units. What is the area of a circular sector such that the apex angle at the center is 90°?
   (a) 0.5 square unit
   (b) 1.0 square unit
   (c) π square units
   (d) It can't be determined without more information

9. What is the perimeter of the circular sector described in Question 8? Include the line segments connecting the center of the circle to the arc.
   (a) $\pi/2$ units
   (b) π units
   (c) $\pi + 4$ units
   (d) It can't be determined without more information

10. What is the tangent of $\pi/4$ rad? Use a calculator if you need it. Find the value of π according to the method described earlier in this chapter. Determine your answer to four decimal places.
    (a) 0.7071
    (b) 0.0137
    (c) 1.0000
    (d) 0.0000

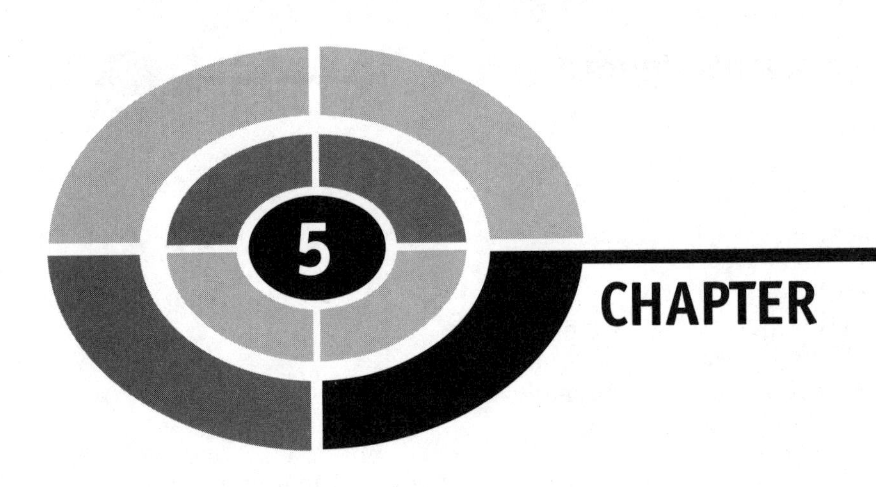

Compass and Straight Edge

In geometry, a *construction* is a drawing made with the simplest possible instruments. Constructions are a powerful learning technique, because they force you to think about the properties of geometric objects, independent of numeric lengths and angle measures. Constructions are also challenging intellectual games.

Tools and Rules

The most common type of geometric construction is done with two instruments, both of which you can purchase at any office supply store. One instrument lets you draw circles, and the other lets you draw straight line segments. Once you have these, you can use them only according to certain "rules of the game."

DRAFTSMAN'S COMPASS

The *draftsman's compass* is a device for drawing circles of various sizes. It has two straight shafts joined at one end with a hinge. One shaft ends in a sharp point that does not mark anything, but that can be stuck into a piece of paper as an anchor. The other shaft has brackets in which a pen or pencil is mounted. To draw a circle, press the sharp point down on a piece of paper (with some cardboard underneath to protect the table or desk top), open the hinge to get the desired radius, and draw the circle by rotating the whole assembly at least once around. You can draw arcs by rotating the compass only partway around.

For geometric constructions, the compass must not have an angle measurement scale at its hinge. If it has a scale that indicates angle measures or otherwise quantifies the extent to which it is opened, you must ignore that scale.

STRAIGHT EDGE

A *straight edge* is any object that helps you to draw line segments by placing a pen or pencil against the object and running it alongside. A conventional ruler will work for this purpose, but is not the best tool to use because it has a calibrated scale. A better tool is a *drafting triangle.* Use any edge of the triangle as the straight edge.

Ignore the angles at the apexes of a drafting triangle. Some drafting triangles have two 45° angles and one 90° angle; others have one 30° angle, one 60° angle, and one 90° angle. You aren't allowed to take advantage of these standard angle measures when performing geometric constructions, so it doesn't matter which type of drafting triangle you use.

WHAT'S ALLOWED

With a compass, you can draw circles or arcs having any radius you want. The center point can be randomly chosen, or you can place the sharp tip of the compass down on a predetermined, existing point and make it the center point of the circle or arc.

You can set a compass to replicate the distance between any two defined points, by setting the non-marking tip down on one point and the marking tip down on the other point, and then holding the compass setting constant.

With the straight edge, you can draw line segments of any length, up to the entire length of the tool. You can draw a random line segment, or choose a specific point through which the line segment passes, or connect any two specific points with a line segment.

WHAT'S NOT ALLOWED

Whatever sort of circle or line segment you draw, you are not allowed to measure the radius or the length against a calibrated scale of any sort. You may not measure angles using a calibrated device. You may not make any reference marks on either the compass or the straight edge. (Marking on a straight edge is "cheating," but referencing a distance using a compass is acceptable, even though the two acts are qualitatively similar!)

Here is a subtle but important restriction: You may not make use of the result of an infinite number of operations (imagining that it is possible), or an infinite number of repetitions of a single operation. That means, for example, that you cannot mentally do a maneuver over and over *ad infinitum* to geometrically approach a desired result, and then claim that result as a valid construction. The entire operation must be completed in a finite number of steps.

DEFINING POINTS

To define an arbitrary point, all you need to do is draw a little dot on the paper. Alternatively, you can set the non-marking point of the compass down on the paper, in preparation for drawing an arc or circle centered at an arbitrary point. Points can also be defined where two line segments intersect, where an arc or circle intersects a line segment, or where an arc or circle intersects another arc or circle.

DRAWING LINE SEGMENTS

Line segments can be drawn in three ways: arbitrarily, through (or starting at) a single point, and through (or connecting) two points.

When you want to draw an arbitrary line segment, place the straight edge down on the paper and run a pencil along the edge (Fig. 5-1A). You can make it as long or as short as you want, but never longer than the length of the straight edge. If you need to draw a line segment longer than the straight edge, don't align the straight edge with part of the line segment and then try to extend it. Use a longer straight edge, so you can create the entire segment in one swipe.

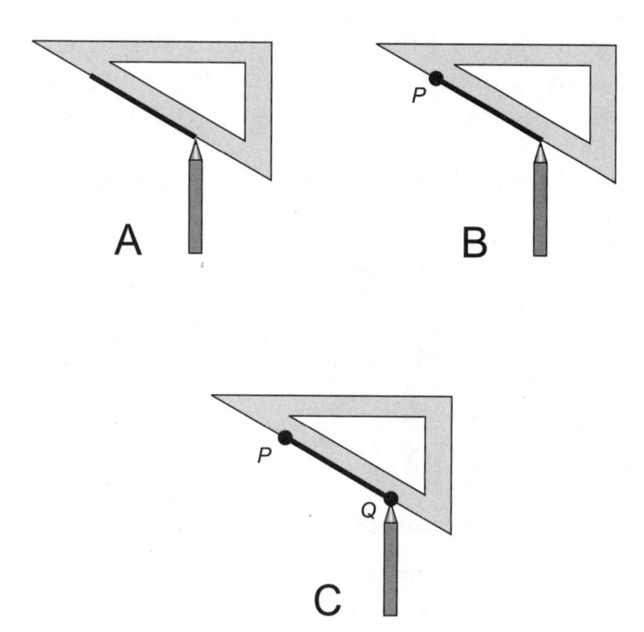

Fig. 5-1. At A, construction of an arbitrary line segment. At B, construction of a line segment starting at a single predetermined point. At C, construction of a line segment connecting two predetermined points.

When you want to draw a line segment through a single defined point, place the tip of the pencil on that point (call it point *P*), place the straight edge down against the point of the pencil, and then run the pencil back and forth along the edge. If you want the point to be an end point of the line segment, run the pencil away from the point in one direction (Fig. 5-1B).

When you want to draw a line segment through two defined points (call them *P* and *Q*), place the tip of the pencil on one of the points, place the straight edge down against the point of the pencil, rotate the straight edge until it lines up with the other point while still firmly resting against the tip of the pencil, and then run the pencil back and forth along the edge, so the mark passes through both points. If you want the points to be the end points of the line segment, make sure the pencil makes its mark only between the points, and not past them on either side (Fig. 5-1C).

DENOTING RAYS

In order to denote a ray, first determine or choose the end point of the ray. Then place the tip of the pencil at the end point, and place the straight edge against the tip of the pencil. Orient the straight edge so it runs in the direction you want the ray to go. Move the tip of the pencil away from the point in the direction of the ray, as far as you want without running off the end of the straight edge (Fig. 5-2A). Finally, draw an arrow at the end of the line segment you have drawn, opposite the starting point (Fig. 5-2B). The arrow indicates that the ray extends infinitely in that direction.

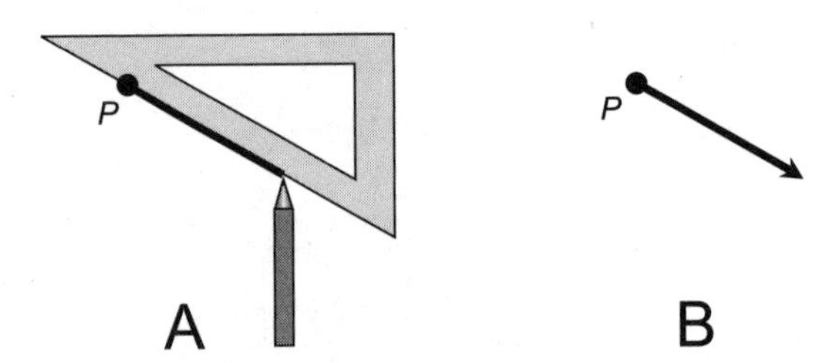

Fig. 5-2. Construction of a ray. First construct a line segment ending at a point (A); then put an arrow at the end opposite the point (B).

DENOTING LINES

In order to draw a line, follow the same procedure as you would to draw a line segment. Then place arrows at both ends (Fig. 5-3). A line can be drawn arbitrarily (as shown at A and B), through a single defined point (as shown at C and D), or through two defined points (as shown at E and F).

DRAWING CIRCLES

To draw a circle around a random point, place the non-marking tip of the compass down on the paper, set the compass to the desired radius, and rotate the instrument through a full circle (Fig. 5-4A). If the center point is predetermined (marked by a dot), place the non-marking tip down on the dot and rotate the instrument through a full circle.

DRAWING ARCS

To draw an arc centered at a random point, place the non-marking tip of the compass down on the paper, set the compass to the desired radius, and rotate the instrument through the desired arc. If the center point is predetermined (marked by a dot), place the non-marking tip down on the dot and rotate the instrument through the desired arc (Fig. 5-4B).

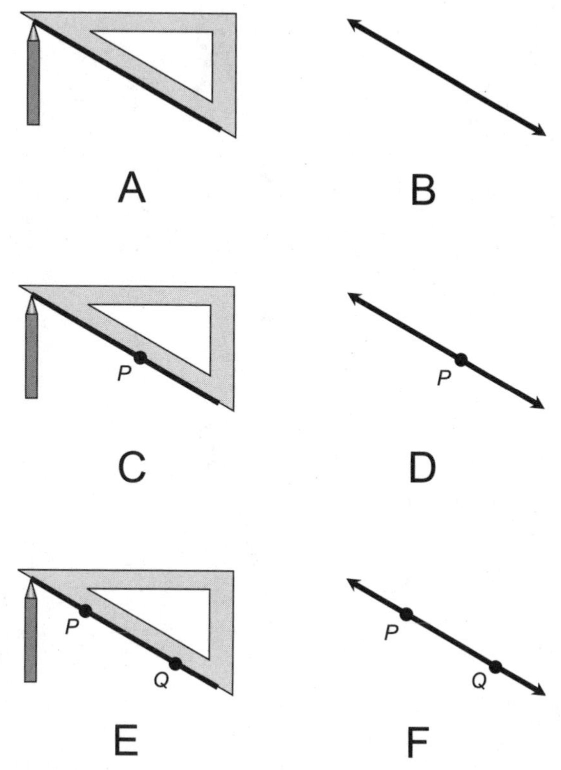

Fig. 5-3. At A and B, construction of an arbitrary line. At C and D, construction of a line through a single predetermined point. At E and F, construction of a line through two predetermined points.

PROBLEM 5-1

Define a point by drawing a dot. Then, with the compass, draw a small circle centered on the dot. Now construct a second circle, concentric with the first one, but having twice the radius.

SOLUTION 5-1

Figure 5-5 illustrates the procedure. In drawing A, the circle is constructed with the compass, centered at the initial point (called point P). In drawing B, a line segment L is drawn using the straight edge, with one end at point P and passing through the circle at a point Q. The line segment extends outside the circle for a distance considerably greater than the circle's radius. In drawing C, a circle is constructed, centered at point Q and leaving the compass set for the same radius as it was when the original circle was drawn. This new circle intersects L at point P (the center of the original circle) and also at a new point R. Next, the non-marking tip of the compass is placed back at point P,

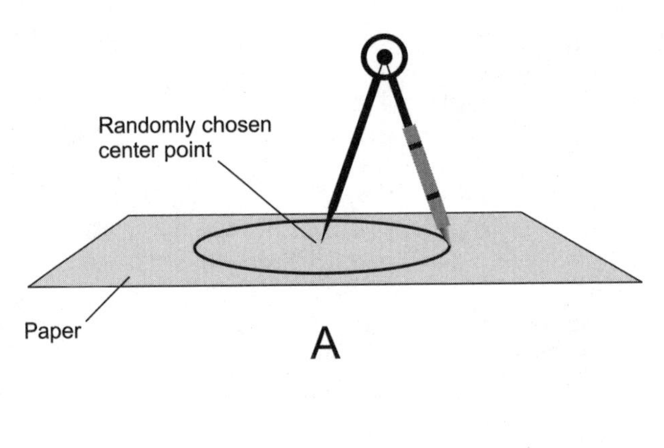

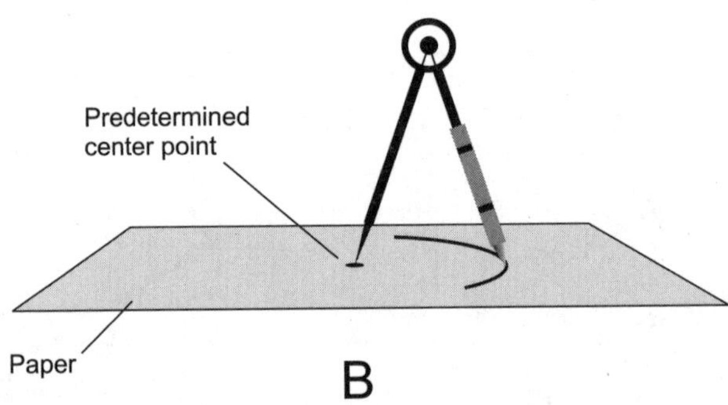

Fig. 5-4. At A, a compass is used to draw a circle around a randomly chosen center point. At B, a compass is used to draw an arc centered at a predetermined point.

and the compass is opened up so the pencil tip falls exactly on point *R*. Finally, as shown in drawing D, a new circle is drawn with its center at point *P*, with a radius equal to the length of line segment *PR*.

PROBLEM 5-2

Draw three points on a piece of paper, placed so they do not all lie along the same line. Label the points *P*, *Q*, and *R*. Construct ΔPQR connecting these three points. Draw a circle whose radius is equal to the length of side *PQ*, but that is centered at point *R*.

SOLUTION 5-2

The process is shown in Fig. 5-6. In drawing A, the three points are put down and labeled. In drawing B, the points are connected to form ΔPQR. Drawing C shows how the non-marking tip of the compass is placed at point *Q*, and the tip of the pencil is placed on point *P*. (You don't have to draw the arc, but it is included in this illustration for emphasis.) With the compass thus set so it defines the length of line segment *PQ*, the non-

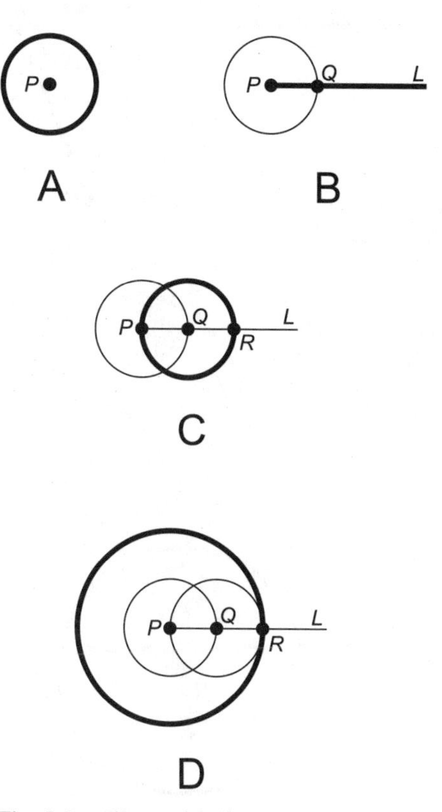

Fig. 5-5. Illustration for Problem 5-1.

marking tip of the compass is placed on point R. Finally, as shown in drawing D the circle is constructed.

PROBLEM 5-3

Can the non-marking tip of the compass be placed at point P, and the pencil tip placed to draw an arc through point Q, in order to define the length of line segment PQ in Problem 5-2?

SOLUTION 5-3

Yes. This will work just as well.

Linear Constructions

The following paragraphs describe how to perform various constructions with line segments. By extension, these same processes apply to rays and lines; you can extend line segments and add arrows as necessary.

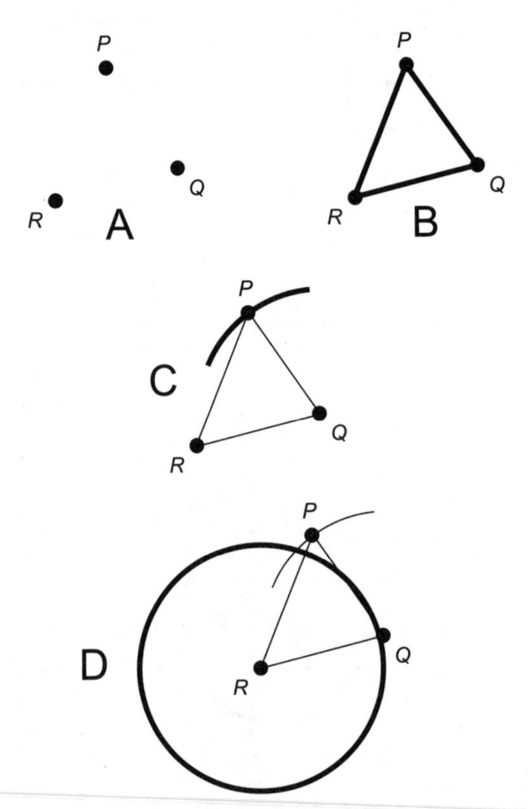

Fig. 5-6. Illustration for Problems 5-2 and 5-3.

REPRODUCING A LINE SEGMENT

Suppose you have a line segment whose end points are P and Q (as shown in Fig. 5-7A), and you want to create another line segment having the same length as PQ. First, construct a "working segment" that is somewhat longer than PQ. Then place a point on this "working segment" and call it R, as shown in drawing B. Next, take the compass and set down the non-marking tip on point P, and adjust the compass spread so the tip of the pencil lands exactly on point Q. By doing this, you have defined the length of line segment PQ using the compass.

Next, place the non-marking tip of the compass down on point R, and create a small arc that intersects your "working segment," as shown in drawing C. Define the intersection of the "working segment" and the arc as point S. The length of line segment RS is the same as that of PQ, so you have reproduced line segment PQ (drawing D).

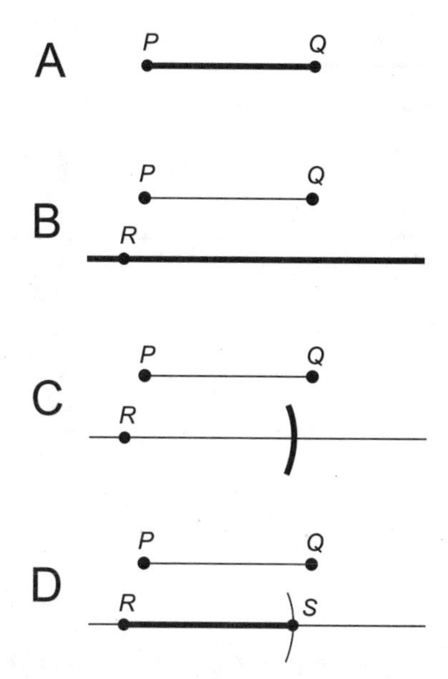

Fig. 5-7. Reproducing a line segment.

BISECTING A LINE SEGMENT

Suppose you have a line segment PQ (Fig. 5-8A) and you want to find the point at its center, that is, the point that bisects line segment PQ. First, construct an arc centered at point P. Make the arc roughly half-circular, and set the compass to span somewhat more than half the length of PQ. Then, without altering the setting of the compass, draw an arc centered at point Q, such that its radius is the same as that of the first arc you drew (as shown at B). Name the points at which the two arcs intersect R and S. Construct a line passing through both R and S. Line RS intersects the original line segment PQ at a point T, which bisects line segment PQ (as shown at C).

PERPENDICULAR BISECTOR

Suppose you have a line segment PQ and you want to construct a line that bisects PQ, and that also passes perpendicularly through PQ. Figure 5-8 shows how this line (called RS in this example) is constructed as a byproduct of the bisection process.

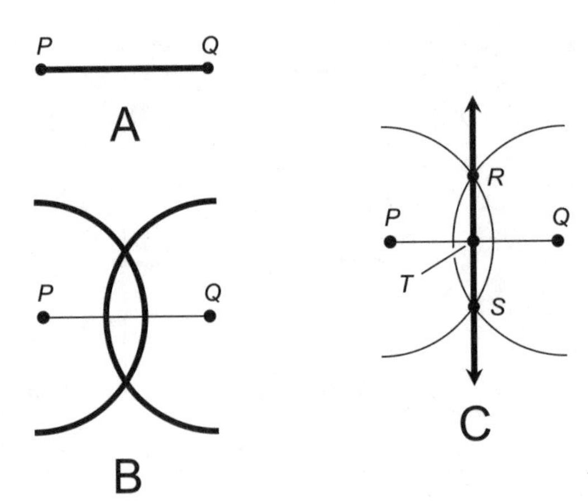

Fig. 5-8. Bisecting a line segment, and constructing a perpendicular bisector.

PERPENDICULAR RAY AT DEFINED POINT

Figure 5-9 illustrates the construction of a perpendicular ray from a defined point *P* on a line or line segment.

Begin with the scenario at drawing A. Set the compass for a moderate span, and construct two arcs opposite each other, both centered at point *P*, that intersect the line or line segment. Call these intersection points *Q* and *R*, as shown in drawing B. Increase the span of the compass, roughly doubling it. Construct an arc centered at *Q* and another arc centered at *R*, so the two arcs have the same radius and intersect as shown in drawing C. Call this intersection point *S*. Construct a ray whose initial point is *P*, and that passes through *S*. Ray *PS* is perpendicular to the original line or line segment at the original defined point *P*.

DROPPING A PERPENDICULAR TO A LINE

Figure 5-10 shows how to drop a perpendicular from a defined point *P* to a line nearby. The term *dropping a perpendicular* means that a line segment, line, or ray is constructed through a point, such that it "comes down on" the nearby line at a right angle.

Begin with the situation shown at A. Set the compass for a span somewhat greater than the distance between *P* and the line. Construct an arc that passes through the line at two points. Call these points *Q* and *R*, as shown in drawing B.

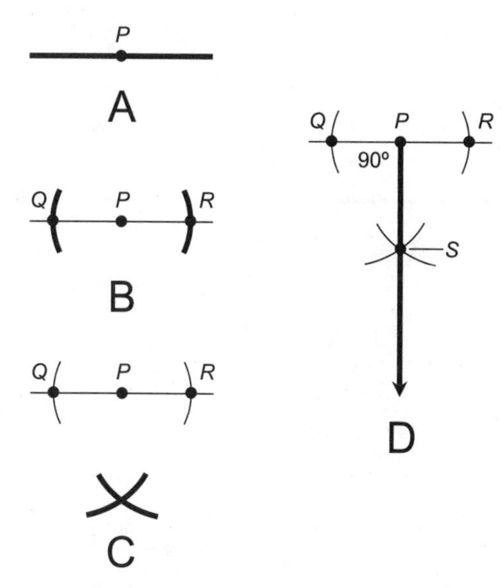

Fig. 5-9. Constructing a ray perpendicular to a line or line segment.

Now increase the span of the compass, roughly doubling it. Construct two arcs, one centered at point Q and the other centered at point R, such that the two arcs have the same radius and intersect each other (drawing C). Call this intersection point S. Construct a line segment that passes through point S and the original defined point P, and extend this line segment until it intersects the original line. Call this intersection point T. Line segment PT intersects the original line at a right angle; that is, PT is a perpendicular "dropped" from point P to the original line.

PARALLEL TO A LINE THROUGH A SPECIFIC POINT

There are several ways to construct a parallel to a line through a specific point that does not lie on that line. One of these methods takes advantage of previous constructions, and is shown in Fig. 5-11.

Suppose you have a line segment with a point P nearby (as shown at A), and you want to create a line through P parallel to the original line. First, drop a perpendicular from P to the line using the procedure described above and shown in Fig. 5-10, generating points Q, R, S, and T (Fig. 5-11B). Then set the compass for the distance PT, and construct a circle centered at P having a radius equal to the distance PT. This gives rise to a new point, U, on line PT, where the circle intersects line PT. Line segment UP has the same length as line segment PT (drawing C).

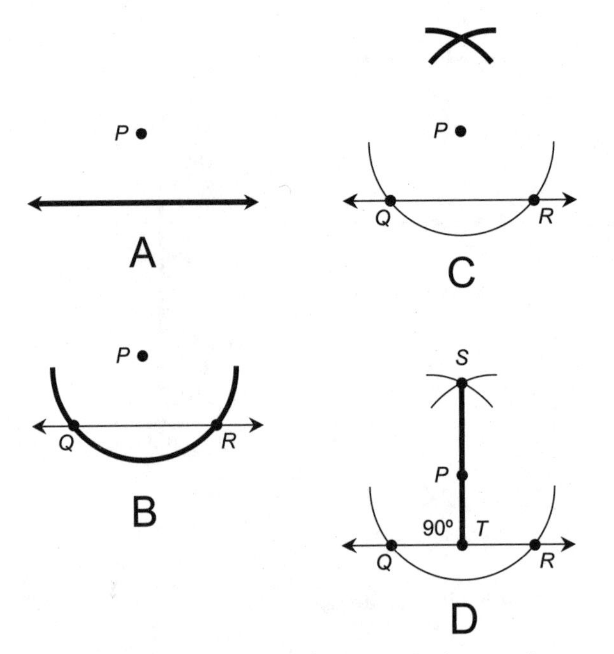

Fig. 5-10. Dropping a perpendicular to a line.

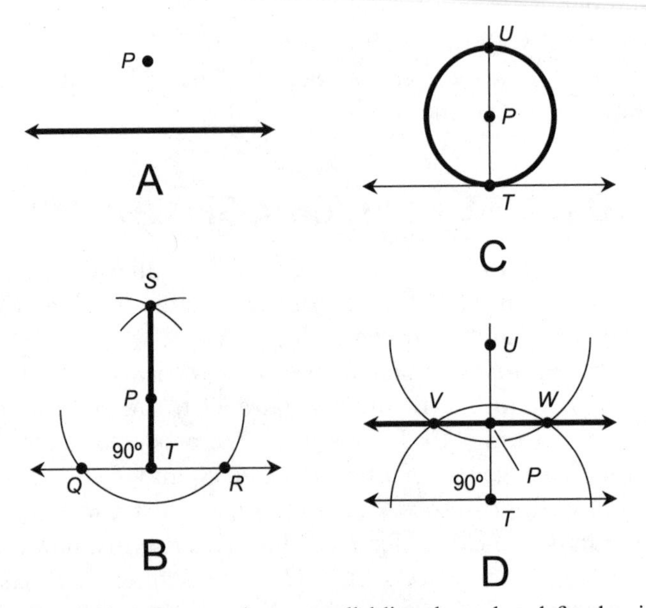

Fig. 5-11. Constructing a parallel line through a defined point.

Now increase the span of the compass somewhat, and construct two roughly half-circular arcs having identical radii, one centered at point T and the other centered at point U, so the arcs intersect each other at two new points, V and W. Line VW is perpendicular to line UT and also to line PT. (We know this because we have just performed the perpendicular construction described earlier.) But PT is perpendicular to the original line. Therefore, line VW is parallel to the original line. This is an example in which we can rightly say "Two perpendiculars make a parallel."

PROBLEM 5-4

Find another way to construct a parallel to a line through a given point not on that line.

SOLUTION 5-4

The initial situation is shown in Fig. 5-12A. The following method is one possible example; there may be others.

First, drop a perpendicular from point P to the original line, as described earlier in this chapter. This intersects the line at point Q (Fig. 5-12B). Next, set the compass so its span is equal to the length of line segment PQ. You can set the non-marking point of the compass down on point Q, and draw an arc through P to be sure you get the compass span just right.

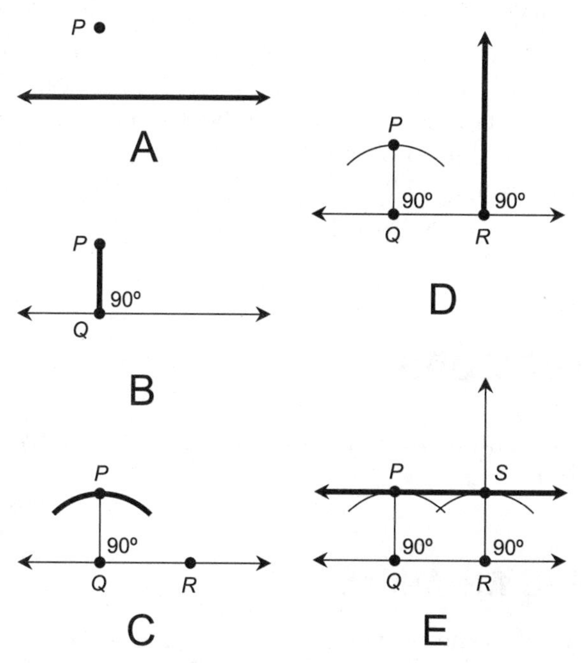

Fig. 5-12. Illustration for Problems 5-4 and 5-5.

Now choose a second point, R, on the line (as shown in drawing C). Construct a perpendicular ray at point R according to the procedure described earlier in this chapter (drawing D). Set the non-marking point of the compass down on point R, and draw an arc that intersects the ray. Call the intersection point S. Now you have two points, P and S, that are equidistant from the original line. Construct line PS through these points. Line PS is parallel to the original line (as shown at E).

PROBLEM 5-5

Construct a square. It doesn't have to be any particular size, as long as all four sides are the same length and all four interior angles measure 90° ($\pi/2$ rad).

SOLUTION 5-5

Examine Fig. 5-12. The quadrilateral $PQRS$ is a rectangle. We know this because line segments PQ and RS are both perpendicular to line QR, so both $\angle PQR$ and $\angle QRS$ are right angles. We also know that lines QR and PS are parallel because that is the intended outcome of Problem 5-4. Therefore it follows that $\angle RSP$ and $\angle SPQ$ are right angles, because opposite interior angles to the transversals of parallel lines always have equal measure; they are congruent ($\angle PQR \cong \angle RSP$ and $\angle QRS \cong \angle SPQ$).

Knowing this, it is a short step to modify the construction process shown in Fig. 5-12 to ensure that the resulting quadrilateral $PQRS$ is a square. Instead of choosing point R on the original line at random, use the compass, set so its span is equal to the distance PQ, to determine point R. Set the non-marking point of the compass down on point Q, and draw an arc so that it intersects the original line to obtain point R. This ensures that the distance PQ is equal to the distance QR. From there, complete the construction in the same way as was done to solve Problem 5-4.

Angular Constructions

The following paragraphs describe how to reproduce (copy) an angle, and also how to bisect an angle.

REPRODUCING AN ANGLE

Figure 5-13 illustrates the process for reproducing an angle. First, suppose two rays intersect at a point P, as shown in drawing A. Set down the non-

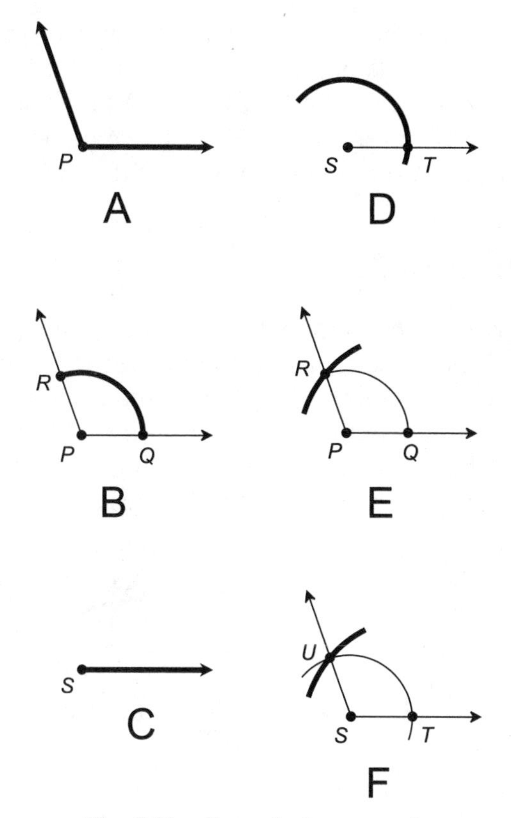

Fig. 5-13. Reproducing an angle.

marking tip of the compass on point *P*, and construct an arc from one ray to the other. Let *R* and *Q* be the two points where the arc intersects the rays (drawing B). Call the angle in question $\angle RPQ$, where points *R* and *Q* are equidistant from point *P*.

Now, place a new point *S* somewhere on the page a good distance away from point *P*, and construct a ray emanating outward from point *S*, as shown in illustration C. (This ray can be in any direction, but it's easiest if you make it go in approximately the same direction as ray *PQ*.) Make the new ray at least as long as ray *PQ*. Without changing the compass span from its previous setting, place its non-marking tip down on point *S* and construct a sweeping arc that is larger than arc *QR*. (You can do this by estimation, as shown in drawing D. You can make a full circle if you want.) Let point *T* represent the intersection of the new arc and the new ray.

Now return to the original arc, place the non-marking tip of the compass down on point *Q*, and construct a small arc through point *R* so the compass spans the distance QR, as shown in drawing E. Then, without changing the span of the compass, place its non-marking tip on point T, and construct an arc that intersects the arc centered on point S. Call this intersection point U. Finally, construct ray SU, as shown in drawing F. You now have a new angle with the same measure as the original angle. That is, $\angle UST \cong \angle RPQ$.

BISECTING AN ANGLE

Figure 5-14 illustrates one method that can be used to bisect an angle, that is, to divide it in half. First, suppose two rays intersect at a point P, as shown in drawing A. Set down the non-marking tip of the compass on point P, and construct an arc from one ray to the other. Call the two points where the arc intersects the rays point R and point Q (drawing B). We can now call the angle in question $\angle RPQ$, where points R and Q are equidistant from point P.

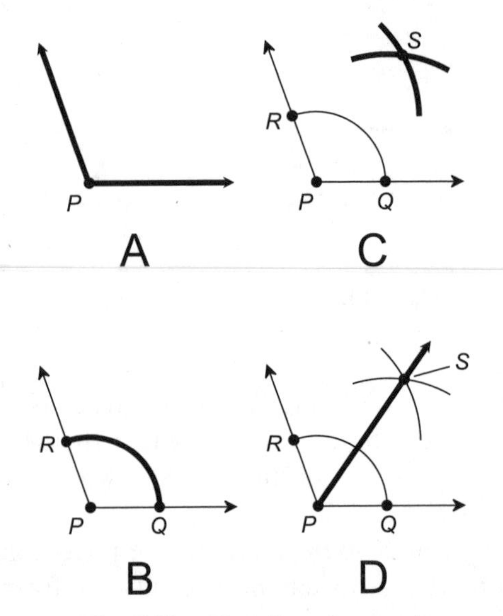

Fig. 5-14. Bisection of an angle.

Now, place the non-marking tip of the compass on point Q, increase its span somewhat from the setting used to generate arc QR, and construct a new arc. Next, without changing the span of the compass, set its non-marking tip down on point R and construct an arc that intersects the arc centered on point Q. (If the arc centered on point Q isn't long enough, go back and make it longer. You can make it a full circle if you want.) Let S be the point at which the two arcs intersect (drawing C). Finally, construct ray PS, as shown at D. This ray bisects $\angle RPQ$. This means that $\angle RPS \cong \angle SPQ$, and also that the sum of the measures of $\angle RPS$ and $\angle SPQ$ is equal to the measure of $\angle RPQ$.

PROBLEM 5-6

Find another way to bisect an angle.

SOLUTION 5-6

Refer to Fig. 5-15. The process starts in the same way as described above. Two rays intersect at point *P*, as shown in drawing A. Set down the non-marking tip of the compass on point *P*, and construct an arc from one ray to the other to get points *R* and *Q* (drawing B) defining $\angle RPQ$, where points *R* and *Q* are equidistant from point *P*.

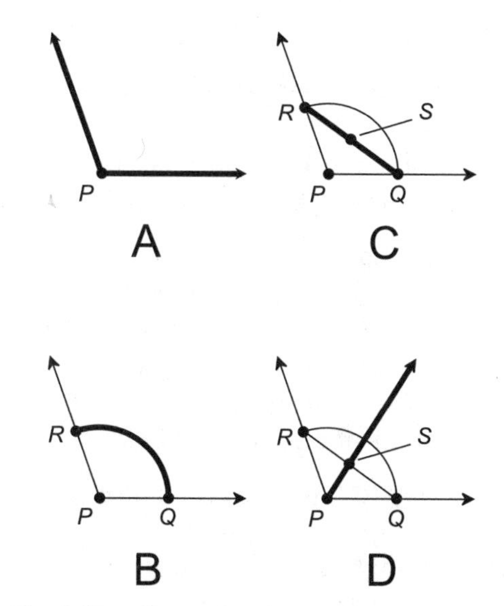

Fig. 5-15. Illustration for Problems 5-6 and 5-7.

Construct line segment *RQ*. Then bisect it, according to the procedure for bisecting line segments described earlier in this chapter. Call the midpoint of the line segment point *S*, as shown in drawing C. Finally, construct ray *PS* (drawing D). This ray bisects $\angle RPQ$.

PROBLEM 5-7

Prove that the angle bisection method described in Solution 5-6 really works.

SOLUTION 5-7

Examine Fig. 5-15D, and note the two triangles $\triangle SRP$ and $\triangle PQS$. These triangles have corresponding sides that have equal lengths:

- $SR = SQ$ (we bisected the line segment)
- $RP = QP$ (we constructed them both from the same arc centered at point P)
- $PS = PS$ (this is trivial)

From these three facts, we know from Chapter 2 that $\triangle SRP$ and $\triangle PQS$ are inversely congruent. This means that corresponding angles (the angles opposite corresponding sides), as we proceed around the triangles in opposite directions, have equal measure. Thus, because $RS = QS$, we can conclude that $\angle RPS$ and $\angle SPQ$ have equal measure. Because their measures obviously add up to the measure of $\angle RPQ$, this proves that ray PS bisects $\angle RPQ$, that is, it cuts the angle in half.

Quiz

Refer to the text in this chapter if necessary. A good score is eight correct. Answers are in the back of the book.

1. Examine Fig. 5-12E. Based only on the information shown in this drawing, which of the following statements can we be certain is true?
   (a) Quadrilateral $PSRQ$ is a parallelogram but not a rectangle
   (b) Quadrilateral $PSRQ$ is a rhombus but not a parallelogram
   (c) Quadrilateral $PSRQ$ is a rectangle
   (d) We can't be certain of any of the above

2. The most obvious way to "quadrisect" a large angle (that is, to divide it into four angles of equal measure) is to
   (a) bisect the large angle using two different construction schemes, and then bisect the resulting angle
   (b) bisect the large angle, and then bisect each of the smaller angles resulting from the bisection
   (c) construct a line segment connecting the rays defining the large angle, bisect the line segment, bisect each of the smaller line segments resulting, and then construct rays from the angle apex through each point generated by the bisections
   (d) give up, because there is no way to "quadrisect" an angle

3. A large angle can be "trisected" (divided into three angles of equal measure) by
   (a) drawing an arc centered at the angle vertex, and then trisecting the arc

(b) drawing an arc centered at the angle vertex, then drawing two arcs centered at the resulting points on the rays defining the angle, and finally drawing rays connecting the points at which the arcs intersect each other
(c) drawing an arc centered at the angle vertex, then drawing a line segment connecting the points at which the arc intersects the rays defining the angle, and finally trisecting the line segment
(d) none of the above means

4. Which of the following operations (a), (b), or (c) is not a "legal" thing to do when performing a construction?
(a) Drawing a circle around a specified point
(b) Drawing a circle around a randomly chosen point
(c) Drawing a straight line through two specified points
(d) All of the above operations (a), (b), and (c) are "legal"

5. Suppose you draw an arbitrary line and an arbitrary point P near that line. Then, using a compass, you construct a circle centered at point P, making the circle large enough so that it intersects the line in two points Q and R. The points P, Q, and R lie at the vertices of
(a) a right triangle
(b) an equilateral triangle
(c) an isosceles triangle
(d) none of the above

6. Suppose you want to construct a trapezoid. The exact measurement of the interior angles or side lengths is not important. The only thing that matters is that the final figure be a true trapezoid. The easiest way to start is to
(a) construct two parallel lines
(b) construct two perpendicular lines
(c) construct a circle
(d) construct two concentric circles

7. A pencil and straight edge cannot be used all by themselves to
(a) construct an arbitrary line
(b) connect two existing, specified points with a line segment
(c) copy an existing, specified line segment
(d) construct an arbitrary angle

8. In Fig. 5-15, the fact that $\triangle SRP$ and $\triangle PQS$ are inversely congruent means that they
(a) are exact mirror images of each other

(b) are the same size, and one can be laid down over the other simply by moving and rotating one of them
(c) are different sizes, but have corresponding interior angle measures that are in the same proportions
(d) are the same size, but have corresponding side lengths that might be in different proportions

9. A compass and a pencil cannot be used all by themselves to
(a) construct an arbitrary circle
(b) construct two line segments of the same length on an existing, specified line
(c) construct two concentric circles around an existing, specified point
(d) construct a straight line segment

10. Suppose you want to construct an angle whose measure is 45°. You could do this by
(a) constructing a square and then drawing its diagonal
(b) constructing a perpendicular bisector line to an existing line segment, and then bisecting one of the angles at which they intersect
(c) bisecting a 180° angle, and then bisecting one of the resulting angles
(d) any of the above methods (a), (b), or (c)

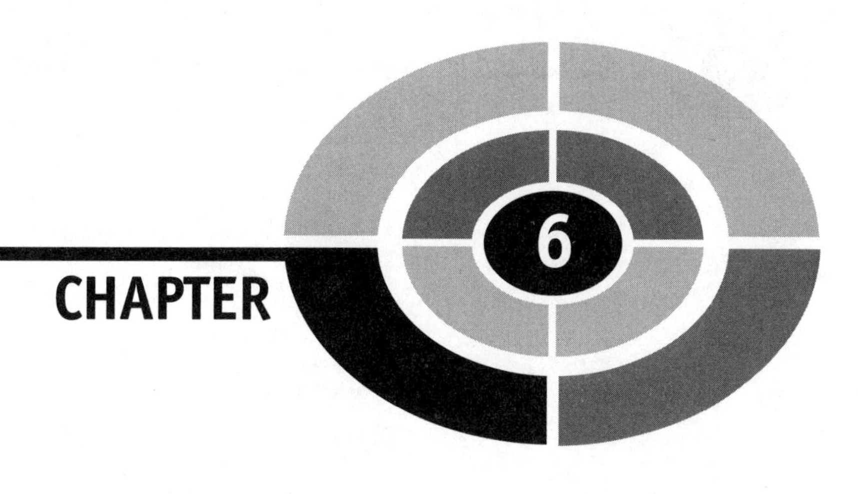

The Cartesian Plane

The *Cartesian plane*, also called the *rectangular coordinate plane* or *rectangular coordinates*, is defined by two number lines that intersect at a right angle. This makes it possible to pictorially render equations that relate one variable to another. You should have a knowledge of middle-school algebra before tackling this chapter. Upon casual observation, some of the equations in this chapter look a little complicated, but nothing here goes beyond middle-school algebra.

Two Number Lines

Figure 6-1 illustrates the simplest possible set of rectangular coordinates. Both number lines have equal increments. This means that on either axis, points corresponding to consecutive integers are the same distance apart, no matter where on the axis we look. The two number lines intersect at their zero points. The horizontal (right-and-left) axis is called the *x axis*; the vertical (up-and-down) axis is called the *y axis*.

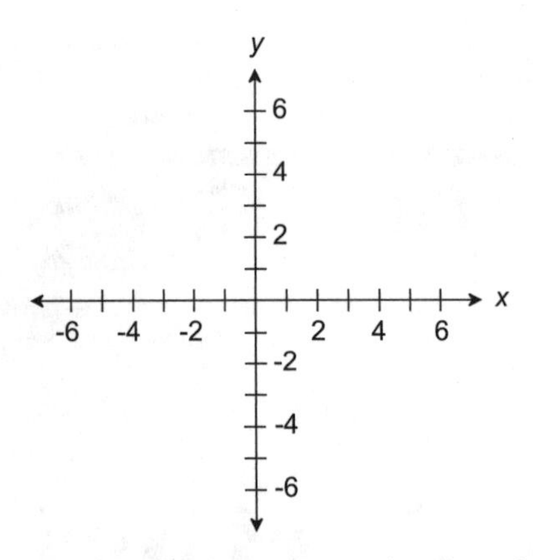

Fig. 6-1. The Cartesian plane is defined by two number lines that intersect at right angles.

ORDERED PAIRS AS POINTS

Figure 6-2 shows two specific points, called P and Q, plotted on the Cartesian plane. The coordinates of point P are $(-5, -4)$, and the coordinates of point Q are $(3,5)$. Any given point on the plane can be denoted as an *ordered pair* in the form (x,y), determined by the numerical values at which perpendiculars from the point intersect the x and y axes. In Fig. 6-2, the perpendiculars are shown as horizontal and vertical dashed lines. When denoting an ordered

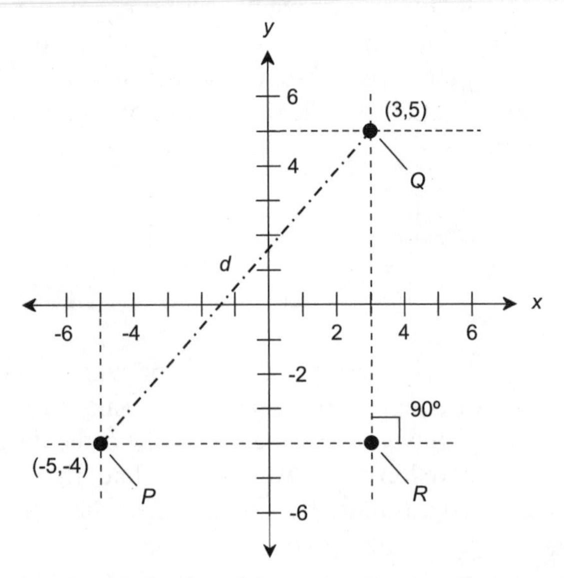

Fig. 6-2. Two points P and Q, plotted in rectangular coordinates, and a third point R, important in finding the distance d between P and Q.

pair, it is customary to place the two numbers or variables together right up against the comma. There is no space after the comma.

The word "ordered" means that the order in which the numbers are listed is important. The ordered pair (7,2) is not the same as the ordered pair (2,7), even though both pairs contain the same two numbers. In this respect, ordered pairs are different than mere sets of numbers. Think of a highway, which consists of a northbound lane and a southbound lane. If there is never any traffic on the highway, it doesn't matter which lane (the one on the eastern side or the one on the western side) is called "northbound" and which is called "southbound." But when there are cars and trucks on that road, it makes a big difference! The untraveled road is like a set; the traveled road is like an ordered pair.

ABSCISSA, ORDINATE, AND ORIGIN

In any graphing scheme, there is at least one *independent variable* and at least one *dependent variable*. As the name suggests, the value of the independent variable does not "depend" on anything; it just "happens." The value of the dependent variable is affected by the value of the independent variable.

The independent-variable coordinate (usually x) of a point on the Cartesian plane is called the *abscissa*, and the dependent-variable coordinate (usually y) is called the *ordinate*. The point (0,0) is called the *origin*. In Fig. 6-2, point P has an abscissa of –5 and an ordinate of –4, and point Q has an abscissa of 3 and an ordinate of 5.

DISTANCE BETWEEN POINTS

Suppose there are two different points $P = (x_0,y_0)$ and $Q = (x_1,y_1)$ on the Cartesian plane. The distance d between these two points can be found by determining the length of the hypotenuse, or longest side, of a right triangle PQR, where point R is the intersection of a "horizontal" line through P and a "vertical" line through Q. In this case, "horizontal" means "parallel to the x axis," and "vertical" means "parallel to the y axis." An example is shown in Fig. 6-2. Alternatively, we can use a "horizontal" line through Q and a "vertical" line through P to get the point R. The resulting right triangle in this case has the same hypotenuse, line segment PQ, as the triangle determined the other way.

Think back to Chapter 2, and recall the Pythagorean theorem. It states that the square of the length of the hypotenuse of a right triangle is equal to the sum of the squares of the other two sides. In this case, that means:

$$d^2 = (x_1 - x_0)^2 + (y_1 - y_0)^2$$

and therefore:

$$d = [(x_1 - x_0)^2 + (y_1 - y_0)^2]^{1/2}$$

where the $1/2$ power is the square root. In the situation shown in Fig. 6-2, the distance d between points $P = (x_0,y_0) = (-5,-4)$ and $Q = (x_1,y_1) = (3,5)$ is:

$$\begin{aligned} d &= \{[3 - (-5)]^2 + [5 - (-4)]^2\}^{1/2} \\ &= [(3+5)^2 + (5+4)^2]^{1/2} \\ &= (8^2 + 9^2)^{1/2} \\ &= (64 + 81)^{1/2} \\ &= 145^{1/2} \\ &= 12.04 \text{ (approx.)} \end{aligned}$$

This is accurate to two decimal places, as determined using a standard digital calculator that can find square roots.

Relation versus Function

It's important to know the similarities and differences between two concepts as they pertain to coordinate geometry: the idea of a *relation* and the idea of a *function*. A relation is an equation or formula that relates the value of one variable to that of another. A function is a relation that meets certain requirements.

RELATIONS

Mathematical relations between two variables x and y are often written in such a way that y is expressed in terms of x. When this is done, y is the dependent variable and x is the independent variable. The following are some examples of relations denoted this way:

$$y = 5$$
$$y = x + 1$$
$$y = 2x$$
$$y = x^2$$

SOME SIMPLE GRAPHS

Figure 6-3 shows how the graphs of the above equations look on the Cartesian plane. Mathematicians and scientists call such a graph a *curve*, even if it happens to be a straight line.

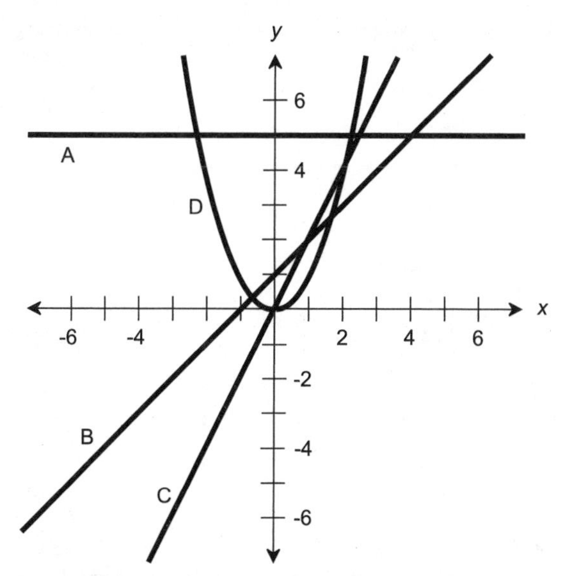

Fig. 6-3. Graphs of four relations in Cartesian coordinates. The relations shown by A, B, and C are linear, but the relation shown by D is not.

The graph of $y = 5$ (curve A) is a horizontal line passing through the point (0,5) on the y axis. The graph of $y = x + 1$ (curve B) is a straight line that ramps upward at a 45° angle (from left to right) and passes through (0,1) on the y axis. The graph of $y = 2x$ (curve C) is a straight line that ramps upward more steeply, and that passes through the origin (0,0). The graph of $y = x^2$ (curve D) is known as a *parabola*. In this case the parabola rests on the origin (0,0), opens upward, and is symmetrical with respect to the y axis.

RELATIONS VS FUNCTIONS

All of the relations graphed in Fig. 6-3 have something in common. For every abscissa, each relation contains at most one ordinate. Never does a curve have two or more ordinates for a single abscissa, although one of them (the parabola, curve D) has two abscissas for all positive ordinates.

A function is a mathematical relation in which every abscissa corresponds to at most one ordinate. According to this criterion, all the curves shown in Fig. 6-3 are graphs of functions of y in terms of x. In addition, curves A, B, and C show functions of x in terms of y. But curve D does not represent a function of x in terms of y. If x is considered the dependent variable, then there are some values of y (that is, some abscissas) for which there exist two values of x (ordinates).

Functions are denoted as italicized letters of the alphabet, usually f, F, g, G, h, or H, followed by the independent variable or variables in parentheses. Examples are:

$$f(x) = x + 1$$
$$g(y) = 2y$$
$$H(z) = z^2$$

These equations are read "f of x equals x plus 1," "g of y equals $2y$," and "H of z equals z squared," respectively.

PROBLEM 6-1

Plot the following points on the Cartesian plane: (–2,3), (3,–1), (0,5), and (–3,–3).

SOLUTION 6-1

These points are shown in Fig. 6-4. The dashed lines are perpendiculars, dropped to the axes to show the x and y coordinates of each point. (The dashed lines are not parts of the coordinates themselves.)

PROBLEM 6-2

What is the distance between (0,5) and (–3,–3) in Fig. 6-4? Express the answer to three decimal places.

SOLUTION 6-2

Use the distance formula. Let $(x_0,y_0) = (0,5)$ and $(x_1,y_1) = (-3,-3)$. Then:

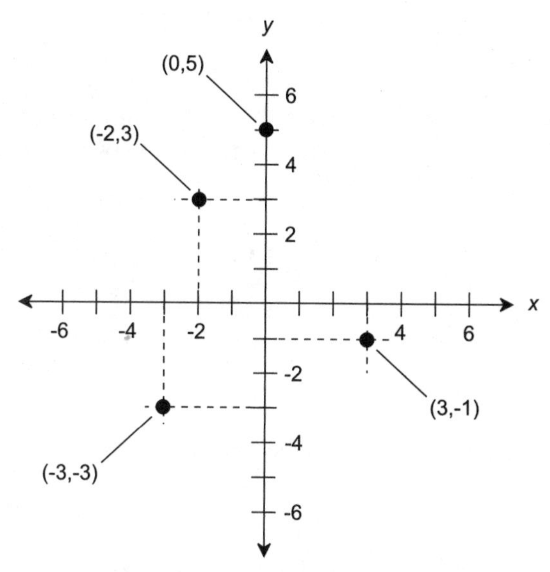

Fig. 6-4. Illustration for Problems 6-1 and 6-2.

$$\begin{aligned} d &= [(x_1 - x_0)^2 + (y_1 - y_0)^2]^{1/2} \\ &= [(-3 - 0)^2 + (-3 - 5)^2]^{1/2} \\ &= [(-3)^2 + (-8)^2]^{1/2} \\ &= (9 + 64)^{1/2} \\ &= 73^{1/2} \\ &= 8.544 \text{ (approx.)} \end{aligned}$$

Straight Lines

Straight lines on the Cartesian plane are represented by a certain type of equation called a *linear equation.* There are several forms in which a linear equation can be written. All linear equations can be reduced to a form where neither x nor y is raised to any power other than 0 or 1.

STANDARD FORM OF LINEAR EQUATION

The standard form of a linear equation in variables x and y consists of constant multiples of the two variables, plus another constant, all summed up to equal zero:

$$ax + by + c = 0$$

In this equation, the constants are a, b, and c. If a constant happens to be equal to 0, then it is not written down, nor is its multiple (by either x or y) written down. Examples of linear equations in the standard form are:

$$\begin{aligned} 2x + 5y - 3 &= 0 \\ 5y - 3 &= 0 \\ 2x - 3 &= 0 \\ 2x &= 0 \\ 5y &= 0 \end{aligned}$$

The last two of these equations can be simplified to $x = 0$ and $y = 0$, by dividing each side by 2 and 5, respectively.

SLOPE–INTERCEPT FORM OF LINEAR EQUATION

A linear equation in variables x and y can be manipulated so it is in a form that is easy to plot on the Cartesian plane. Here is how a linear equation in standard form can be converted to *slope–intercept form*:

$$\begin{aligned} ax + by + c &= 0 \\ ax + by &= -c \\ by &= -ax - c \\ y &= (-a/b)x - c/b \\ y &= (-a/b)x + (-c/b) \end{aligned}$$

where a, b, and c are real-number constants, and $b \neq 0$. The quantity $-a/b$ is called the *slope* of the line, an indicator of how steeply and in what sense the line slants. The quantity $-c/b$ represents the ordinate (or y-value) of the point at which the line crosses the y axis; this is called the *y-intercept*.

WHAT IS SLOPE?

Let dx represent a small change in the value of x on such a graph; let dy represent the change in the value of y that results from this change in x. The ratio dy/dx is the slope of the line, and is symbolized m. Let k represent the y-intercept. Then m and k can be derived from the coefficients a, b, and c as follows, provided $b \neq 0$:

$$m = -a/b$$
$$k = -c/b$$

The linear equation can be rewritten in slope–intercept form as:

$$y = (-a/b)x + (-c/b)$$

and therefore:

$$y = mx + k$$

To plot a graph of a linear equation in Cartesian coordinates, proceed as follows:

- Convert the equation to slope–intercept form
- Plot the point $y = k$
- Move to the right by n units on the graph
- If m is positive, move upward mn units
- If m is negative, move downward $|m|n$ units, where $|m|$ is the absolute value of m
- If $m = 0$, don't move up or down at all
- Plot the resulting point $y = mn + k$
- Connect the two points with a straight line

Figures 6-5A and 6-5B illustrate the following linear equations as graphed in slope–intercept form:

$$y = 5x - 3$$
$$y = -x + 2$$

A positive slope indicates that the line ramps upward as you move from left to right, and a negative slope indicates that the line ramps downward as you move from left to right. A slope of 0 indicates a horizontal line. The slope of a vertical line is undefined because, in the form shown here, it requires that m be defined as a quotient in which the denominator is equal to 0.

POINT–SLOPE FORM OF LINEAR EQUATION

It is difficult to plot a graph of a line based on the y-intercept (the point at which the line intersects the y axis) when the part of the graph of interest is far from the y axis. In this sort of situation, the *point–slope form* of a linear equation can be used. This form is based on the slope m of the line and the coordinates of a known point (x_0, y_0):

$$y - y_0 = m(x - x_0)$$

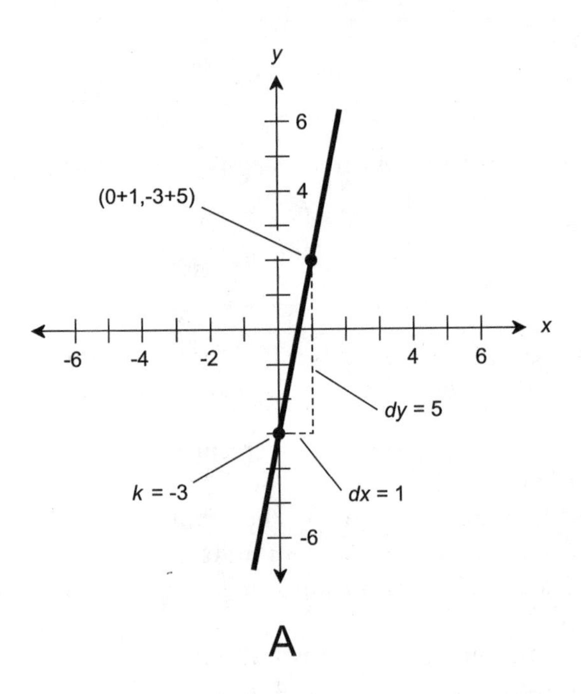

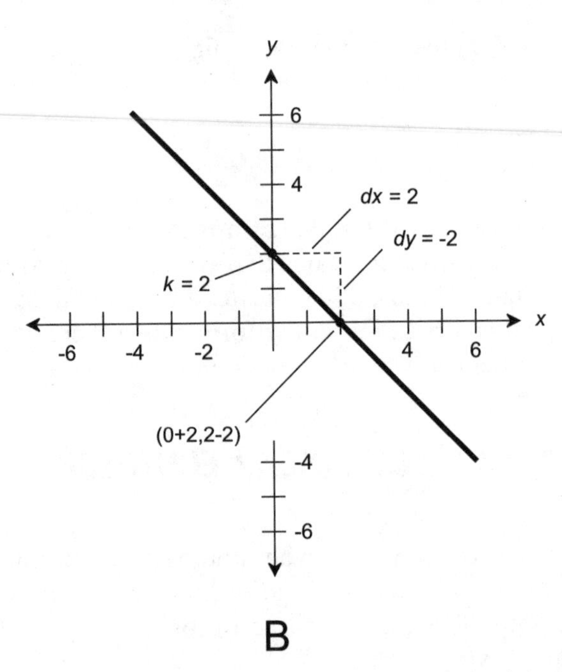

Fig. 6-5. (A) Graph of the linear equation $y = 5x - 3$. (B) Graph of the linear equation $y = -x + 2$.

To plot a graph of a linear equation using the point–slope method, you can follow these steps in order:

- Convert the equation to point–slope form
- Determine a point (x_0, y_0) by "plugging in" values
- Plot (x_0, y_0) on the coordinate plane
- Move to the right by n units on the graph, where n is some number that represents a reasonable distance on the graph
- If m is positive, move upward mn units
- If m is negative, move downward $|m|n$ units, where $|m|$ is the absolute value of m
- If $m = 0$, don't move up or down at all
- Plot the resulting point (x_1, y_1)
- Connect the points (x_0, y_0) and (x_1, y_1) with a straight line

Figure 6-6A illustrates the following linear equation as graphed in point–slope form:

$$y - 104 = 3(x - 72)$$

Figure 6-6B is another graph of a linear equation in point–slope form:

$$y + 55 = -2(x + 85)$$

FINDING LINEAR EQUATION BASED ON GRAPH

Suppose we are working in the Cartesian plane, and we know the exact coordinates of two points P and Q. These two points define a unique and distinct straight line. Call the line L. Let's give the coordinates of the points these names:

$$P = (x_p, y_p)$$
$$Q = (x_q, y_q)$$

The slope m of line L can be found using either of the following formulas:

$$m = (y_q - y_p)/(x_q - x_p)$$
$$m = (y_p - y_q)/(x_p - x_q)$$

provided x_p is not equal to x_q. The point–slope equation of L can be determined based on the known coordinates of P or Q. Therefore, either of the following formulas represent the line L:

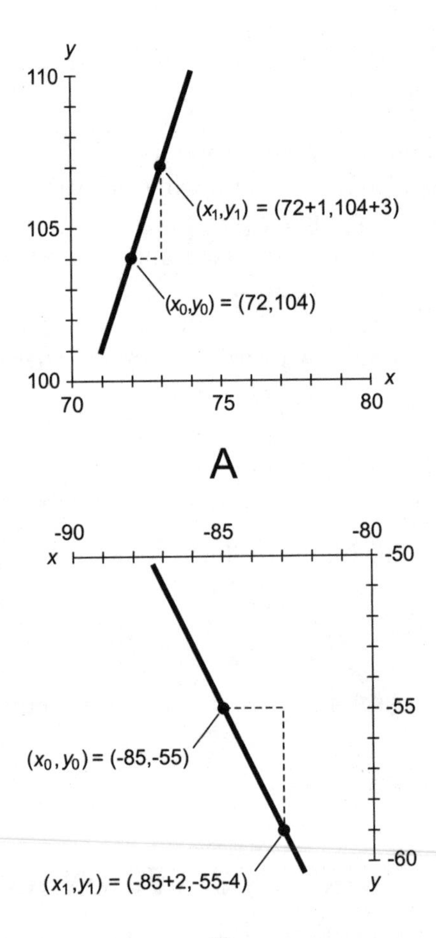

Fig. 6-6. (A) Graph of the linear equation $y - 104 = 3(x - 72)$. (B) Graph of the linear equation $y + 55 = -2(x + 85)$.

$$y - y_p = m(x - x_p)$$
$$y - y_q = m(x - x_q)$$

Parabolas and Circles

The Cartesian-coordinate graph of a *quadratic equation* is a *parabola.* In Cartesian coordinates, a quadratic equation looks like this:

$$y = ax^2 + bx + c$$

where a, b, and c are real-number constants, and $a \neq 0$. (If $a = 0$, then the equation is linear, not quadratic.) To plot a graph of an equation that appears in the above form, first determine the coordinates of the following point (x_0,y_0):

$$x_0 = -b/(2a)$$
$$y_0 = c - b^2/(4a)$$

This point represents the *base point* of the parabola; that is, the point at which the curvature is sharpest, and at which the slope of a line tangent to the curve is zero. Once this point is known, find four more points by "plugging in" values of x somewhat greater than and less than x_0, and then determining the corresponding y-values. These x-values, call them x_{-2}, x_{-1}, x_1, and x_2, should be equally spaced on either side of x_0, such that:

$$x_{-2} < x_{-1} < x_0 < x_1 < x_2$$
$$x_{-1} - x_{-2} = x_0 - x_{-1} = x_1 - x_0 = x_2 - x_1$$

This will give five points that lie along the parabola, and that are symmetrical relative to the axis of the curve. The graph can then be inferred (that means we make an educated guess!) if the points are wisely chosen. Some trial and error might be required. If $a > 0$, the parabola opens upward. If $a < 0$, the parabola opens downward.

Let's go ahead and try this with a concrete example.

PLOTTING A PARABOLA

Consider the following formula:

$$y = x^2 + 2x + 1$$

Using the above formula to calculate the base point:

$$x_0 = -2/2 = -1$$
$$y_0 = 1 - 4/4 = 1 - 1 = 0$$

Therefore,

$$(x_0,y_0) = (-1,0)$$

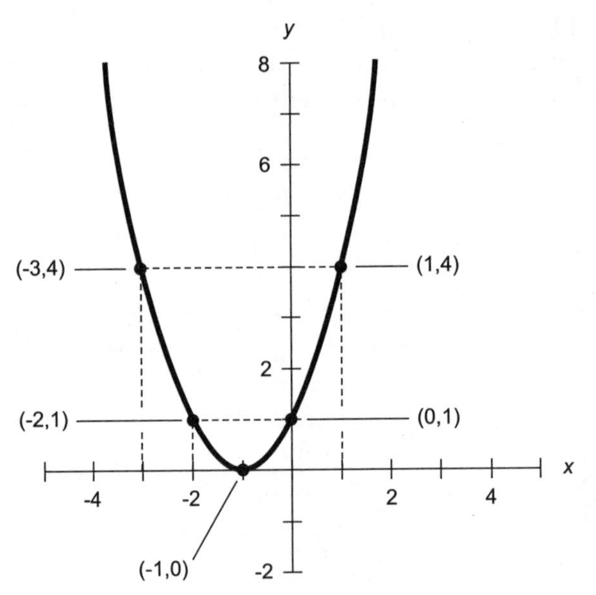

Fig. 6-7. Graph of the quadratic equation $y = x^2 + 2x + 1$.

This point is plotted first, as shown in Fig. 6-7. Next, plot the points corresponding to x_{-2}, x_{-1}, x_1, and x_2, spaced at 1-unit intervals on either side of x_0, as follows:

$$x_{-2} = x_0 - 2 = -3$$
$$y_{-2} = (-3)^2 + 2 \times (-3) + 1 = 9 - 6 + 1 = 4$$

Therefore,

$$(x_{-2}, y_{-2}) = (-3{,}4)$$

$$x_{-1} = x_0 - 1 = -2$$
$$y_{-1} = (-2)^2 + 2 \times (-2) + 1 = 4 - 4 + 1 = 1$$

Therefore,

$$(x_{-1}, y_{-1}) = (-2{,}1)$$

$$x_1 = x_0 + 1 = 0$$
$$y_1 = 0^2 + 2 \times 0 + 1 = 0 + 0 + 1 = 1$$

Therefore,

$$(x_1,y_1) = (0,1)$$

$$x_2 = x_0 + 2 = 1$$

$$y_2 = 1^2 + 2 \times 1 + 1 = 1 + 2 + 1 = 4$$

Therefore,

$$(x_2,y_2) = (1,4)$$

From these five points, the curve can be inferred.

PLOTTING ANOTHER PARABOLA

Let's try another example, this time with a parabola that opens downward. Consider the following formula:

$$y = -2x^2 + 4x - 5$$

The base point is:

$$x_0 = -4/-4 = 1$$

$$y_0 = -5 - 16/(-8) = -5 + 2 = -3$$

Therefore,

$$(x_0,y_0) = (1,-3)$$

This point is plotted first, as shown in Fig. 6-8. Next, plot the following points:

$$x_{-2} = x_0 - 2 = -1$$

$$y_{-2} = -2 \times (-1)^2 + 4 \times (-1) - 5 = -2 - 4 - 5 = -11$$

Therefore,

$$(x_{-2},y_{-2}) = (-1,-11)$$

$$x_{-1} = x_0 - 1 = 0$$

$$y_{-1} = -2 \times 0^2 + 4 \times 0 - 5 = -5$$

Therefore,

$$(x_{-1},y_{-1}) = (0,-5)$$

$$x_1 = x_0 + 1 = 2$$

$$y_1 = -2 \times 2^2 + 4 \times 2 - 5 = -8 + 8 - 5 = -5$$

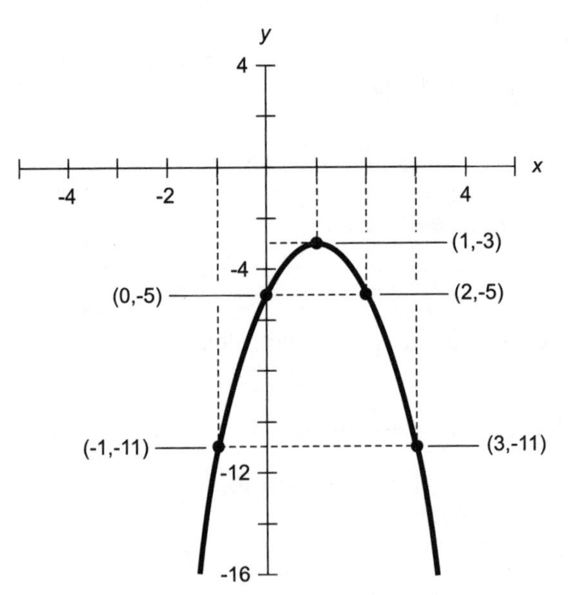

Fig. 6-8. Graph of the quadratic equation $y = -2x^2 + 4x - 5$.

Therefore,

$$(x_1,y_1) = (2, -5)$$

$$x_2 = x_0 + 2 = 3$$

$$y_2 = -2 \times 3^2 + 4 \times 3 - 5 = -18 + 12 - 5 = -11$$

Therefore,

$$(x_2,y_2) = (3, -11)$$

From these five points, the curve can be inferred.

EQUATION OF CIRCLE

The general form for the equation of a *circle* in the *xy*-plane is given by the following formula:

$$(x - x_0)^2 + (y - y_0)^2 = r^2$$

where (x_0,y_0) represents the coordinates of the center of the circle, and r represents the *radius*. This is illustrated in Fig. 6-9. In the special case where the circle is centered at the origin, the formula becomes:

$$x^2 + y^2 = r^2$$

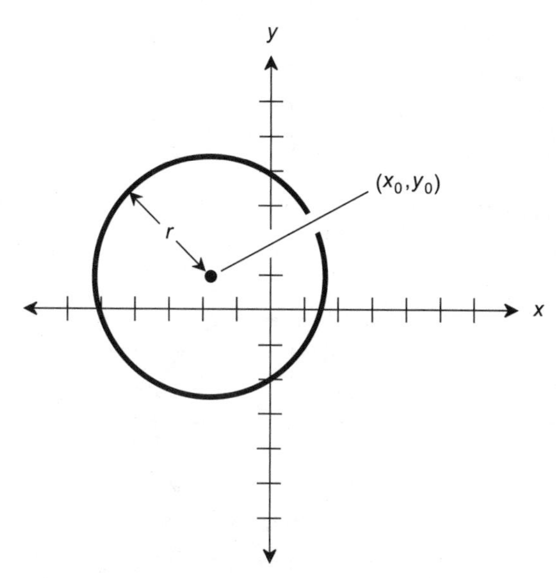

Fig. 6-9. Circle centered at (x_0,y_0) with radius r.

Such a circle intersects the x axis at the points $(r,0)$ and $(-r,0)$; it intersects the y axis at the points $(0,r)$ and $(0,-r)$. An even more specific case is the *unit circle*:

$$x^2 + y^2 = 1$$

This curve intersects the x axis at the points (1,0) and (–1,0); it intersects the y axis at the points (0,1) and (0,–1).

PROBLEM 6-3

Draw a graph of the circle represented by the equation $(x - 1)^2 + (y + 2)^2 = 9$.

SOLUTION 6-3

Based on the general formula for a circle, we can determine that the center point has coordinates $x_0 = 1$ and $y_0 = -2$. The radius is equal to the square root of 9, which is 3. The result is a circle whose center point is (1,–2) and whose radius is 3. This is shown in Fig. 6-10.

PROBLEM 6-4

Determine the equation of the circle shown in Fig. 6-11.

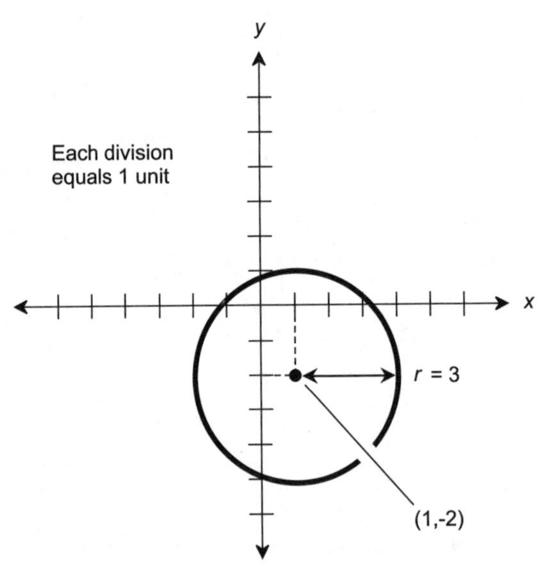

Fig. 6-10. Illustration for Problem 6-3.

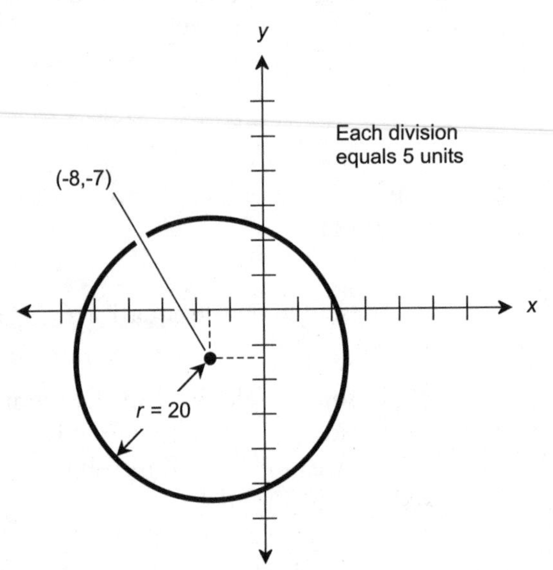

Fig. 6-11. Illustration for Problem 6-4.

SOLUTION 6-4

First note that the center point is (–8,–7). That means $x_0 = -8$ and $y_0 = -7$. The radius, r, is equal to 20, so $r^2 = 20 \times 20 = 400$. Plugging these numbers into the general formula gives us the equation of the circle, as follows:

$$(x - x_0)^2 + (y - y_0)^2 = r^2$$
$$[x - (-8)]^2 + [y - (-7)]^2 = 400$$
$$(x + 8)^2 + (y + 7)^2 = 400$$

Solving Pairs of Equations

The solutions of pairs of equations can be envisioned and approximated by graphing both of the equations on the same set of coordinates. Solutions appear as intersection points between the plots.

A LINE AND A CURVE

Suppose you are given two equations in two variables, such as x and y, and are told to solve for values of x and y that satisfy both equations. Such equations are called *simultaneous equations*. Here is an example:

$$y = x^2 + 2x + 1$$
$$y = -x + 1$$

These equations are graphed in Fig. 6-12. The graph of the first equation is a parabola, and the graph of the second equation is a straight line. The line crosses the parabola at two points, indicating that there are two solutions of this set of simultaneous equations. The coordinates of the points, corresponding to the solutions, can be estimated from the graph. It appears that they are approximately:

$$(x_1,y_1) = (-3,4)$$
$$(x_2,y_2) = (0,1)$$

You can solve the pair of equations using plain algebra, and determine the solutions exactly.

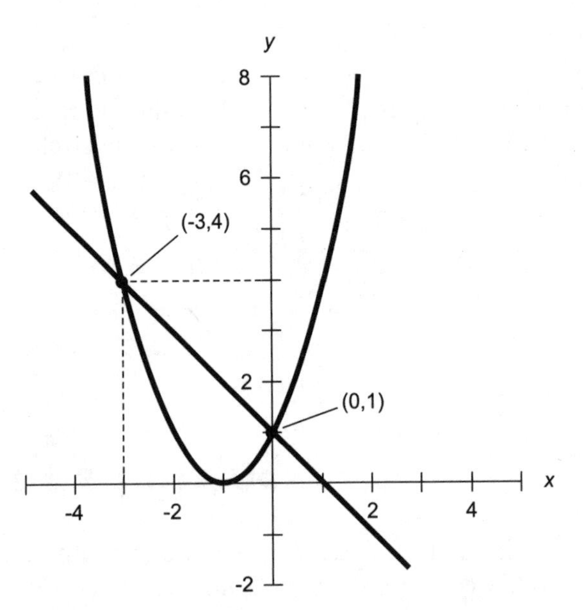

Fig. 6-12. Graphs of two equations, showing solutions as intersection points.

ANOTHER LINE AND CURVE

Here is another pair of "two-by-two" equations (two equations in two variables) that can be approximately solved by graphing:

$$y = -2x^2 + 4x - 5$$
$$y = -2x - 5$$

These equations are graphed in Fig. 6-13. Again, the graph of the first equation is a parabola, and the graph of the second equation is a straight line. The line crosses the parabola at two points, indicating that there are two solutions. The coordinates of the points, corresponding to the solutions, appear to be approximately:

$$(x_1,y_1) = (3, -11)$$
$$(x_2,y_2) = (0, -5)$$

Again, if you want, you can go ahead and solve these equations using algebra, and find the values exactly.

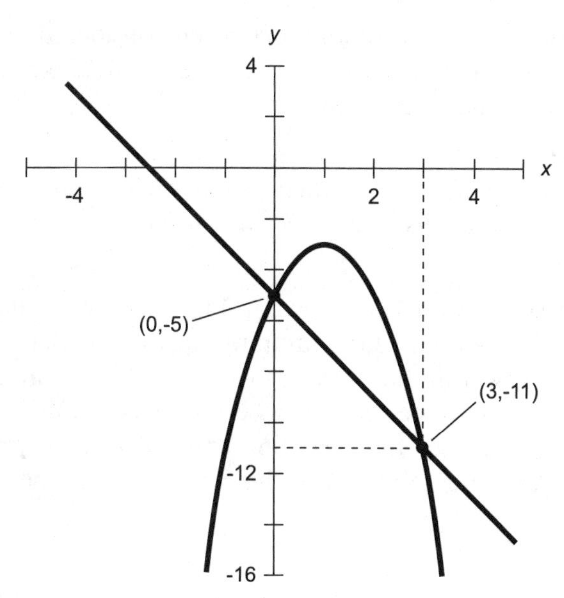

Fig. 6-13. Another example of equation solutions shown as the intersection points of their graphs.

HOW MANY SOLUTIONS?

Graphing simultaneous equations can reveal general things about them, but should not be relied upon to provide exact solutions. In real-life scientific applications, graphs rarely show exact solutions unless they are so labeled and represent theoretical ideals.

A graph can reveal that a pair of equations has two or more solutions, or only one solution, or no solutions at all. Solutions to pairs of equations always show up as intersection points on their graphs. If there are n intersection points between the curves representing two equations, then there are n solutions to the pair of simultaneous equations.

If a pair of equations is complicated, or if the graphs are the results of experiments, you'll occasionally run into situations where you can't use algebra to solve them. Then graphs, with the aid of computer programs to closely approximate the points of intersection between graphs, offer a good means of solving simultaneous equations.

Sometimes you'll want to see if a set of more than two equations in x and y has any solutions shared by them all. It is common for one or more pairs of a large set of equations to have some solutions; this is shown by points where any two of their graphs intersect. But it's unusual for a set of three or more equations in x and y to have any solutions when considered all together (that is, simultaneously). For that to be the case, there must be at least one point that all of the graphs have in common.

PROBLEM 6-5

Using the Cartesian plane to plot their graphs, what can be said about the solutions to the simultaneous equations $y = x + 3$ and $(x - 1)^2 + (y + 2)^2 = 9$?

SOLUTION 6-5

The graphs of these equations are shown in Fig. 6-14. The equation $y = x + 3$ has a graph that is a straight line, ramping up toward the right with slope equal to 1, and intersecting the y axis at (0,3). The equation $(x - 1)^2 + (y + 2)^2 = 9$ has a graph that is a circle whose radius is 3 units, centered at the point (1,–2). It is apparent that this line and circle do not intersect. This means that there exist no solutions to this pair of simultaneous equations.

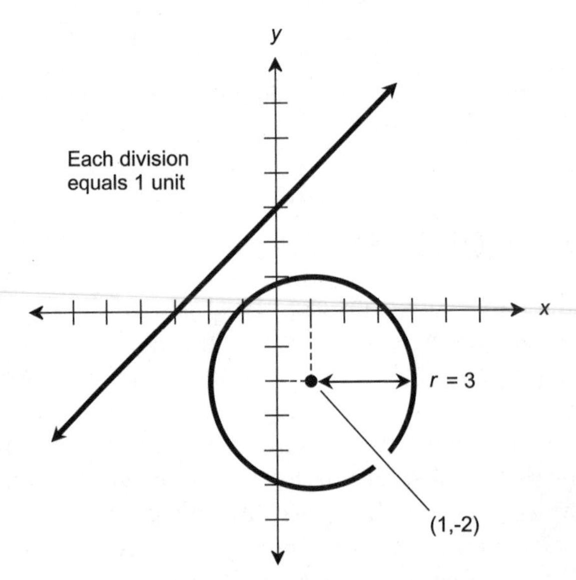

Fig. 6-14. Illustration for Problem 6-5.

PROBLEM 6-6

Using the Cartesian plane to plot their graphs, what can be said about the solutions to the simultaneous equations $y = 1$ and $(x - 1)^2 + (y + 2)^2 = 9$?

SOLUTION 6-6

The graphs of these equations are shown in Fig. 6-15. The equation $y = 1$ has a graph that is a horizontal straight line intersecting the y axis at (0,1). The equation $(x - 1)^2 + (y + 2)^2 = 9$ has a graph that is a circle whose radius is 3

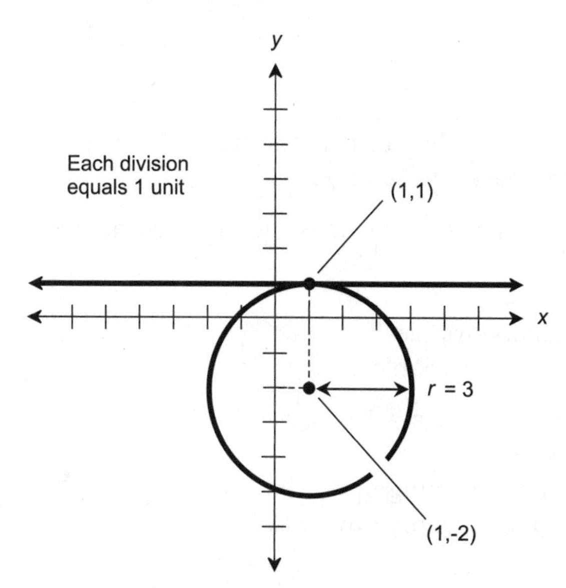

Fig. 6-15. Illustration for Problem 6-6.

units, centered at the point (1,–2). It appears from the graph that the equations have a single common solution denoted by the point (1,1), indicating that $x = 1$ and $y = 1$.

Let's use algebra to solve the equations and see if the graph tells us the true story. Substituting 1 for y in the equation of a circle (because one of the equations tells us that $y = 1$), we get a single equation in a single variable:

$$(x - 1)^2 + (1 + 2)^2 = 9$$
$$(x - 1)^2 + 3^2 = 9$$
$$(x - 1)^2 + 9 = 9$$
$$(x - 1)^2 = 0$$
$$x - 1 = 0$$
$$x = 1$$

It checks out. There is only one solution to this pair of simultaneous equations, and that is $x = 1$ and $y = 1$, denoted by the point (1,1).

Quiz

Refer to the text in this chapter if necessary. A good score is eight correct. Answers are in the back of the book.

1. The ordinate in the xy-plane is the same as the value of the
   (a) abscissa
   (b) x coordinate
   (c) dependent variable
   (d) independent variable

2. The graph of $y = 3x^2 - 5$ is
   (a) a straight line
   (b) a parabola opening upward
   (c) a parabola opening downward
   (d) a circle

3. Suppose you see a graph of a straight line. The x-intercept point is (4,0) and the y-intercept point is (0,8). What is the equation of this line?
   (a) $y = -2x + 8$
   (b) $y = -4x - 8$
   (c) $y = 4x + 8$
   (d) $(x - 4)^2 + (y - 8)^2 = 0$

4. At which points, if any, do the graphs of $y = 2x + 4$ and $y = 2x - 4$ intersect?
   (a) (0,–2)
   (b) (2,0)
   (c) (0,–2) and (2,0)
   (d) The graphs do not intersect

5. Consider a situation similar to the one encountered in Problem 6-6 and Fig. 6-15. Suppose the equation of the circle is changed to $(x - 1)^2 + (y + 2)^2 = 100$, but the equation of the line remains the same at $y = 1$. What can we say about the solutions to the pair of equations now?
   (a) There are none
   (b) There is one
   (c) There are two
   (d) There are more than two

6. Examine Fig. 6-12. At what point does the straight line intersect the x axis?

(a) (0,1)
(b) (1,0)
(c) (–3,4)
(d) It is impossible to precisely tell without more information

7. What are the y-intercept points, if any, of the circle $(x + 5)^2 + (y + 4)^2 = 1$?
   (a) It is impossible to tell without more information
   (b) (0,5) and (0,4)
   (c) (0,–5) and (0,–4)
   (d) There are none

8. What is the distance d between the points (3,5) and (5,3)?
   (a) $d = 0$
   (b) $d = 2$
   (c) $d = 8^{1/2}$
   (d) $d = 4$

9. What is the slope of the line represented by the equation $y - 2 = 3x + 18$?
   (a) 2
   (b) 3
   (c) 18
   (d) –18

10. Examine Fig. 6-12. Suppose that a third equation is graphed on this Cartesian plane, and its equation is $y = x - 3$. If the equations of the two existing graphs are considered together with this new equation, how many common solutions are there to all three equations considered simultaneously?
    (a) None
    (b) One
    (c) Two
    (d) Three

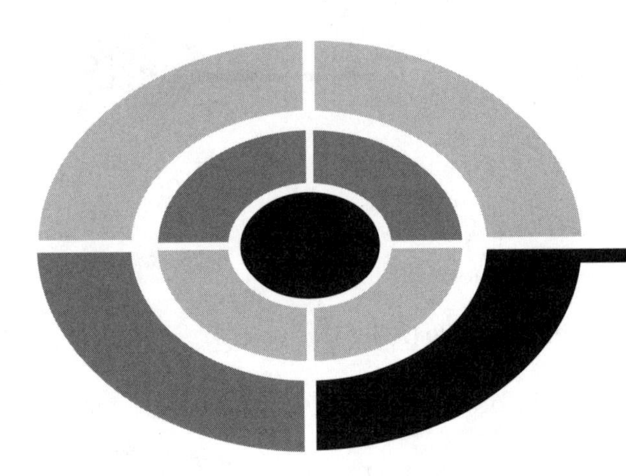

Test: Part One

Do not refer to the text when taking this test. You may draw diagrams or use a calculator if necessary. A good score is at least 38 correct. Answers are in the back of the book. It's best to have a friend check your score the first time, so you won't memorize the answers if you want to take the test again.

1. Two triangles are *directly similar* if and only if
   (a) they are both equilateral
   (b) they are both isosceles
   (c) they have corresponding sides of identical lengths
   (d) they have the same proportions in the same rotational sense
   (e) the sum of the measures of their interior angles is equal to 180°

2. A regular polygon
   (a) has sides that are all the same length, but interior angles whose measures may differ
   (b) has interior angles that all have the same measure, but sides whose lengths may differ
   (c) has sides that are all the same length, and interior angles whose measures are all the same
   (d) has vertices that all lie along the same line
   (e) has interior angles whose measures add up to 360°

3. Which of the following statements is false in Euclidean geometry?
   (a) Two different, parallel, straight lines do not intersect
   (b) A straight line segment has no perpendicular bisectors
   (c) An acute angle measures less than 90°
   (d) Three points, not all on the same line, always lie in the same plane
   (e) Two intersecting straight lines always lie in the same plane

4. Look at Fig. Test 1-1. Suppose that line segments *SQ* and *PR* are perpendicular to each other at their intersection point *T*. From this, we can be certain
   (a) that *PQRS* is a rhombus
   (b) that *PQRS* is a square
   (c) that *PQRS* is a rectangle
   (d) that *PQRS* does not lie in a single plane
   (e) about none of the above

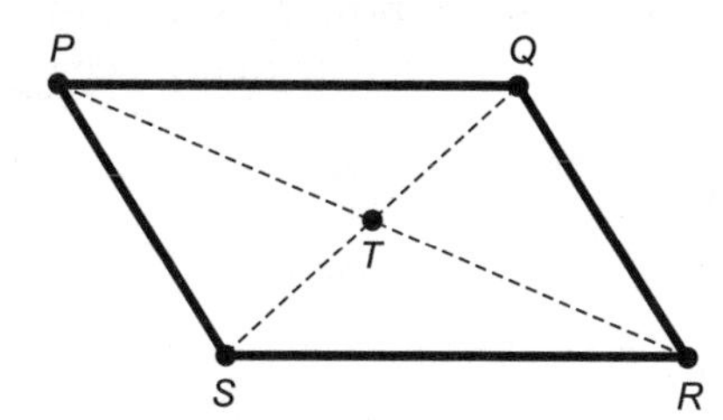

Fig. Test 1-1. Illustration for Questions 4, 5, 6, and 7 in the test for Part One.

5. Suppose that in Fig. Test 1-1, line segments *PQ* and *RS* are parallel, but line segments *PS* and *QR* are not parallel. From this, we can be certain
   (a) that *PQRS* is a rhombus
   (b) that *PQRS* is a square
   (c) that *PQRS* is a rectangle
   (d) that *PQRS* is a parallelogram
   (e) about none of the above

6. Suppose that in Fig. Test 1-1, $\triangle PTQ \cong \triangle RTS$ and $\triangle PTS \cong \triangle RTQ$. From this, we can be certain
   (a) that *PQRS* is a rhombus
   (b) that *PQRS* is a square
   (c) that *PQRS* is a rectangle
   (d) that *PQRS* does not lie in a single plane
   (e) about none of the above

7. Suppose that in Fig. Test 1-1, line segments PS and QR are parallel, and $\triangle PSQ \cong \triangle RQS$. From this, we can be certain
   (a) that $PQRS$ is a rhombus
   (b) that $PQRS$ is a square
   (c) that $PQRS$ is a rectangle
   (d) that $PQRS$ is a parallelogram
   (e) about none of the above

8. When creating a geometric construction, it is important that
   (a) the compass be a precision drafting device, not a dime-store item
   (b) the straight edge be calibrated in metric units
   (c) markings on the straight edge or ruler be non-existent or ignored
   (d) the marking device be a pencil, not a pen
   (e) the paper be white and the marking device be black for maximum contrast

9. How far is the point (10,10) from the origin in Cartesian coordinates? Round off the answer to three decimal places.
   (a) 7.071 units
   (b) 10.000 units
   (c) 14.142 units
   (d) 20.000 units
   (e) 100.000 units

10. A half-open line segment
    (a) has zero length
    (b) contains both of its end points
    (c) contains one, but not both, of its end points
    (d) contains neither of its end points
    (e) has no end points

11. Suppose you see a graph of a circle in Cartesian coordinates. The circle is centered at the origin (that is, the point where $x = 0$ and $y = 0$). The radius of the circle is equal to 10 units. At what point on the circle are the x and y values equal and negative? Round off the values to three decimal places.
    (a) (–10.000,–10.000)
    (b) (–14.142,–14.142)
    (c) (–7.071,–7.071)
    (d) There are infinitely many such points
    (e) There is no such point

12. What is the slope of the graph of the equation $x = 3$ in Cartesian coordinates?
    (a) 3
    (b) –3
    (c) 0
    (d) It is not defined
    (e) More information is needed to determine it

13. Look at Fig. Test 1-2. Suppose that lines *TR*, *QU*, and *QR* are all straight, and that lines *TR* and *QU* are parallel. Which of the following statements is false?
    (a) ∠*TRS* and ∠*RQU* are alternate interior angles
    (b) ∠*TRS* and ∠*PQU* have the same measure
    (c) Line *PR* is a transversal to lines *TR* and *QU*
    (d) ∠*PRT* has the same measure as ∠*RQU*
    (e) All of the above statements are true

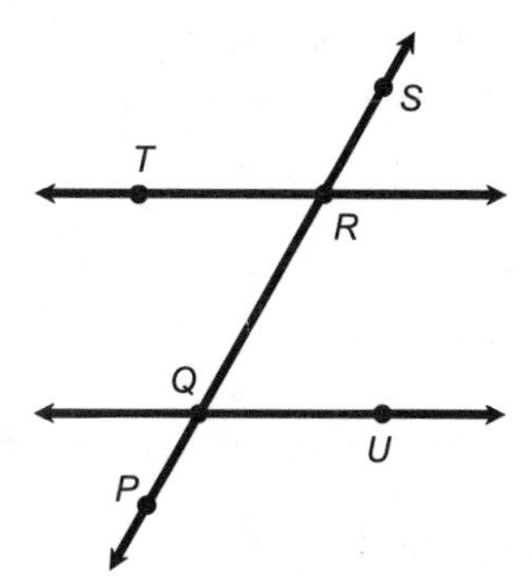

Fig. Test 1-2. Illustration for Questions 13 and 14 in the test for Part One.

14. Look at Fig. Test 1-2. Suppose that lines *TR*, *QU*, and *QR* are all straight and all lie in the same plane. Also suppose that ∠*TRQ* has the same measure as ∠*UQR*. Which of the following statements (a), (b), (c), or (d) is not necessarily true?
    (a) Lines *TR* and *QU* are parallel
    (b) ∠*TRS* has the same measure as ∠*UQP*
    (c) Line *SP* is a transversal to lines *TR* and *QU*
    (d) ∠*QRS* has the same measure as ∠*PQR*
    (e) All of the above statements (a), (b), (c), and (d) are true

15. In a geometric construction, which of the following operations (a), (b), (c), or (d) is not allowed?
    (a) Placing the compass down and adjusting it to span the length of an existing line segment
    (b) Drawing two lines so they intersect at an angle measuring a certain number of degrees as indicated by a protractor
    (c) Drawing a circle of arbitrary radius, centered at an arbitrary point
    (d) Drawing a circle of arbitrary radius, centered at a specific point
    (e) All of the above operations (a), (b), (c), and (d) are allowed

16. If the radius of a circle is 3 units, but nothing else about it is known, then which of the following (a), (b), (c), or (d) cannot be determined?
    (a) The diameter of the circle
    (b) The circumference of the circle
    (c) The perimeter of the circle
    (d) The interior area of the circle
    (e) All of the above (a), (b), (c), and (d) can be determined

17. Suppose a field is shaped like a parallelogram. The long sides measure 100 meters each, and the short sides measure 20 meters each. What is the area of the field?
    (a) 240 square meters
    (b) 200 square meters
    (c) 150 square meters
    (d) 120 square meters
    (e) It is impossible to calculate it without more information

18. Suppose a field is shaped like a parallelogram. The long sides measure 100 meters each, the short sides measure 20 meters each, and the width of the field, as measured at right angles to its long sides, is 15 meters. What is the area of the field?
    (a) 2400 square meters
    (b) 2000 square meters
    (c) 1500 square meters
    (d) 1200 square meters
    (e) It is impossible to calculate it without more information

19. Suppose a Cartesian-coordinate graph shows two straight, parallel lines. How many solutions exist to the pair of simultaneous equations represented by these lines?
    (a) More information is necessary in order to say
    (b) None
    (c) One

(d) Two
(e) Infinitely many

20. Suppose the diagonals of a plane quadrilateral are equally long and they intersect each other at a right angle. Then we can be certain
(a) that the quadrilateral is a square
(b) that the quadrilateral is a rectangle, but not a square
(c) that the quadrilateral is a rhombus, but not a square
(d) that the quadrilateral is a parallelogram, but not a rectangle or a rhombus
(e) about none of the above

21. Suppose the diagonals of a plane quadrilateral are equally long, they intersect each other at their midpoints, and they intersect each other at a right angle. Then we can be certain
(a) that the quadrilateral is a square
(b) that the quadrilateral is a rectangle, but not a square
(c) that the quadrilateral is a rhombus, but not a square
(d) that the quadrilateral is a parallelogram, but not a rectangle or a rhombus
(e) about none of the above

22. A triangle is obtuse if and only if
(a) all its interior angles are obtuse
(b) all its exterior angles are obtuse
(c) one of its interior angles is obtuse
(d) one of its exterior angles is obtuse
(e) the sum of the measures of its interior angles is obtuse

23. In Euclidean geometry, two lines are parallel if and only if
(a) they are in the same plane, and they do not intersect
(b) they are in the same plane, and they intersect at only one point
(c) they are in different planes, and they do not intersect
(d) they are in different planes, and they intersect at only one point
(e) they intersect at a right angle ($90°$ or $\pi/2$ rad)

24. In Fig. Test 1-3, suppose all the sides of the polygon have identical length s (in meters), and all the interior angles have identical measure z (in degrees). This figure is
(a) a regular pentagon
(b) a regular hexagon
(c) a regular septagon
(d) a regular octagon
(e) an irregular polygon

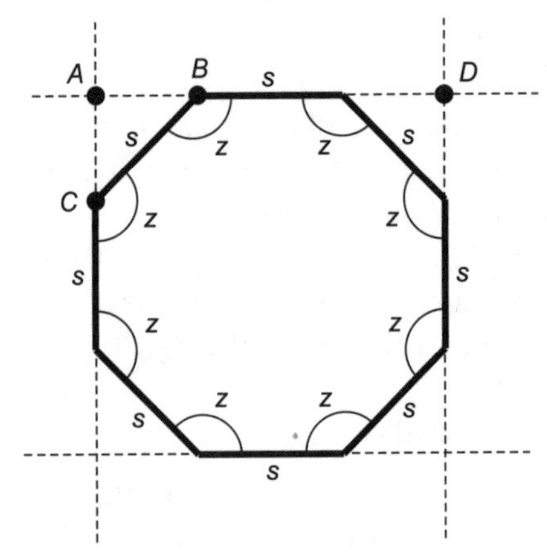

Fig. Test 1-3. Illustration for Questions 24, 25, 26, and 27 in the test for Part One.

25. What is the measure of each interior angle z in the polygon of Fig. Test 1-3, assuming the conditions given in Question 24 all hold?
    (a) 105°
    (b) 120°
    (c) 135°
    (d) 150°
    (e) 165°

26. What is the length of line segment AD in Fig. Test 1-3, assuming the conditions given in Question 24 all hold? (The exponent $^1/_2$ indicates the square root.)
    (a) $s + (2^{1/2}/2)$
    (b) $s + (2^{1/2} \times s)$
    (c) $s + (2 \times 2^{1/2})$
    (d) $2\ (s + 2^{1/2})$
    (e) $2 + s^{1/2}$

27. How can the area of the polygon in Fig. Test 1-3 be found if the value of s is known and all the conditions given in Question 24 hold?
    (a) Determine the area x of the square formed by the dashed lines, then determine the area y of $\triangle ABC$, and finally determine $x + y$
    (b) Determine the area x of the square formed by the dashed lines, then determine the area y of $\triangle ABC$, and finally determine $x + 4y$
    (c) Determine the area x of the square formed by the dashed lines, then determine the area y of $\triangle ABC$, and finally determine $x - y$

(d) Determine the area x of the square formed by the dashed lines, then determine the area y of $\triangle ABC$, and finally determine $x - 4y$
(e) Without more information, this problem cannot be solved

28. Two angles are said to be complementary if and only if
(a) they both have measures of 180°
(b) they are alternate interior angles formed by a transversal to two parallel lines
(c) they are opposite angles formed by the intersection of two lines
(d) the sum of their measures is equal to $\pi/2$ rad
(e) they are equal halves of a bisected angle

29. In a geometric construction, which of the following operations (a), (b), (c), or (d) is allowed?
(a) Drawing a line segment 10 centimeters long, as indicated by a ruler
(b) Drawing an angle measuring 29 degrees, as indicated by a protractor
(c) Drawing a circle with a radius of 5 centimeters, as indicated by a ruler
(d) Using a ruler to draw a straight line through two specific points
(e) None of the above operations (a), (b), (c), or (d) is allowed

30. Imagine two triangles $\triangle ABC$ and $\triangle DEF$. Suppose the names of the vertices of each triangle go alphabetically in order as you proceed counterclockwise. Further suppose that all three of the following hold true:

- Line segment AB is the same length as line segment DE
- $\angle CAB$ has the same measure as $\angle FDE$
- $\angle ABC$ has the same measure as $\angle DEF$

What can we say, with certainty, about $\triangle ABC$ and $\triangle DEF$?

(a) They are directly congruent triangles
(b) They are both isosceles triangles
(c) They are both right triangles
(d) They are both acute triangles
(e) Nothing in particular

31. If the diagonals of a parallelogram are both equally long, then that parallelogram is
(a) a square
(b) a rhombus

(c) a rectangle
(d) an irregular quadrilateral
(e) congruent

32. What is the equation of the straight line in Fig. Test 1-4?
(a) $y = -4x - 1$
(b) $y = 4x - 1$
(c) $y = x - 4$
(d) $y = x + 4$
(e) $y = 4x + 4$

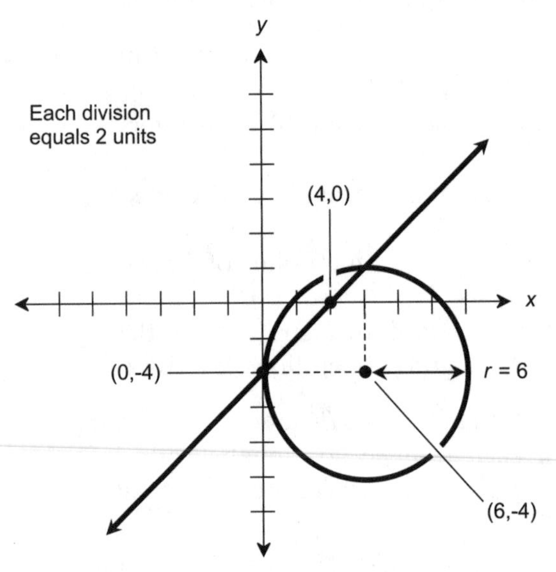

Fig. Test 1-4. Illustration for Questions 32, 33, and 34 in the test for Part One.

33. What is the equation of the circle in Fig. Test 1-4?
(a) $x^2 + y^2 = 6$
(b) $x^2 - y^2 = 6$
(c) $x^2 + y^2 = 36$
(d) $x^2 - y^2 = 36$
(e) None of the above

34. How many solutions exist for the pair of simultaneous equations represented by the line and the circle in Fig. Test 1-4?
(a) None
(b) One
(c) Two

(d) Three
(e) Infinitely many

35. As the number of sides in a regular polygon increases, the interior area of the polygon
(a) increases without limit
(b) approaches the area of a circle inscribed within the polygon
(c) approaches the area of a square inscribed within the polygon
(d) remains constant
(e) becomes undefined

36. Consider a triangle whose vertex points are D, E, and F. Suppose the measure of $\angle DEF$ is equal to $\pi/2$ rad. Further suppose that the length of side DE is 30 meters, and the length of side EF is 40 meters. What is the interior area of $\triangle DEF$?
(a) 1200 square meters
(b) 600 square meters
(c) 500 square meters
(d) 120 square meters
(e) It is impossible to tell without more information

37. Envision again the triangle described in the previous question. What is the perimeter of $\triangle DEF$?
(a) 1200 meters
(b) 600 meters
(c) 500 meters
(d) 120 meters
(e) It is impossible to tell without more information

38. In Fig. Test 1-5, suppose all the points, line segments, lines, and arcs lie in a single plane, arc A has radius u and is centered on point P, and arc B has radius u and is centered on point Q. Which of the following statements logically follows from these facts?
(a) Line segment PT has the same length as line segment RT
(b) Arcs A and B encompass the same angular measure
(c) The length of line segment TQ is equal to u
(d) Line segment PQ has the same length as line segment RS
(e) None of the above

39. In Fig. Test 1-5, suppose line segment PQ is drawn, and then arcs A and B are drawn having equal radii u and centered on points P and Q, respectively. Finally, line RS is drawn through the points at which the arcs intersect each other. This process illustrates a method of

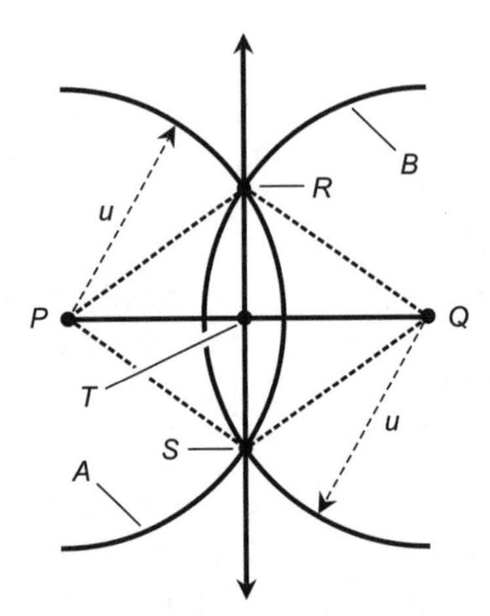

Fig. Test 1-5. Illustration for Questions 38, 39, and 40 in the test for Part One.

(a) angle duplication
(b) arc duplication
(c) line-segment bisection
(d) angle bisection
(e) none of the above

40. In Fig. Test 1-5, suppose all the points, line segments, lines, and arcs lie in a single plane, arc A has radius u and is centered on point P, and arc B has radius u and is centered on point Q. Based on these facts, which of the following statements is not necessarily true?
    (a) Line segment PT has the same length as line segment TQ
    (b) Quadrilateral $PRQS$ is a rhombus
    (c) Line segment RS has the same length as line segment PQ
    (d) $\triangle PTR$ is a right triangle
    (e) All of the above statements are necessarily true

41. An angle of $3\pi/4$ rad has the same measure as an angle of
    (a) 30°
    (b) 45°
    (c) 90°
    (d) 135°
    (e) 180°

42. Suppose a straight line is graphed in Cartesian coordinates. If you move 2 units toward the right (that is, you increase the x-value by 2), the graph moves up by 4 units (that is, the y-value increases by 4). What is the slope of the line?

(a) 2
(b) 4
(c) –2
(d) –4
(e) There is not enough information to tell

43. In a geometric construction, which of the following operations (a), (b), (c), or (d) does not require the use of a drafting compass?
(a) Duplicating a line segment
(b) Bisecting a line segment
(c) Drawing a perpendicular bisector to a line segment
(d) Bisecting an angle
(e) All of the above operations (a), (b), (c), and (d) require the use of a drafting compass

44. In a trapezoid
(a) opposite pairs of angles have equal measure
(b) adjacent pairs of angles have equal measure
(c) the triangles formed by the sides and diagonals are all congruent
(d) the sum of the measures of the interior angles is equal to 360°
(e) the sum of the measures of the interior angles is equal to 180°

45. Suppose a square measures exactly 16 centimeters on a side. What is the perimeter of a circle inscribed within this square?
(a) 4π centimeters
(b) 8π centimeters
(c) 16π centimeters
(d) 32π centimeters
(e) It cannot be determined without more information

46. The graph of the equation $y = 5 - 6x^2$ in Cartesian coordinates is
(a) a straight line with positive slope
(b) a straight line with negative slope
(c) a parabola
(d) a circle
(e) none of the above

47. Suppose there exists a line L in Euclidean geometry. Let P be some point that is not on line L. How many lines can exist that pass through P and are parallel to L?
(a) It is impossible to say without more information
(b) None
(c) One

(d) Two
(e) Infinitely many

48. Suppose, in the triangle shown by Fig. Test 1-6, that the measures of angles x and y are both 60°. Then we can be certain that $\triangle PQR$ is
(a) an equilateral triangle
(b) an isosceles triangle, but not an equilateral triangle
(c) a right triangle

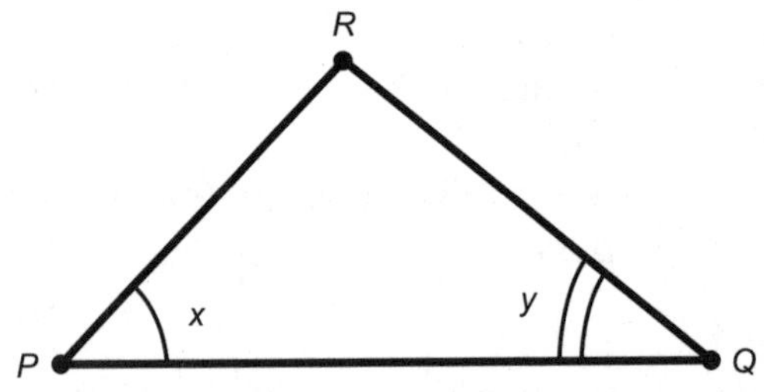

Fig. Test 1-6. Illustration for Questions 48, 49, and 50 in the test for Part One.

(d) a directly congruent triangle
(e) a triangle, but no particular type

49. Examine Fig. Test 1-6. If angles x and y have the same measure, then we can be certain that $\triangle PQR$ is
(a) an equilateral triangle
(b) an isosceles triangle
(c) a right triangle
(d) a directly congruent triangle
(e) a triangle, but no particular type

50. Suppose, in the triangle shown by Fig. Test 1-6, that the sum of the measures of angles x and y is equal to $\pi/2$ rad. Then we can be certain that $\triangle PQR$ is
(a) an equilateral triangle
(b) an isosceles triangle
(c) a right triangle
(d) a directly congruent triangle
(e) a triangle, but no particular type

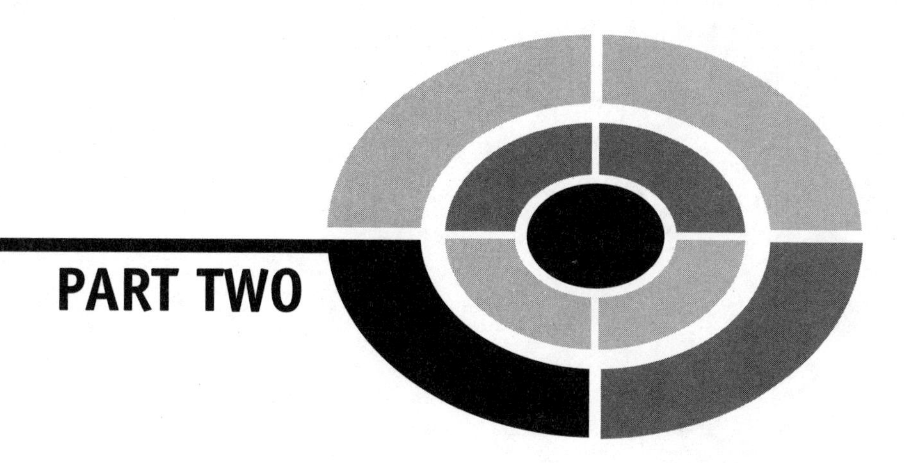

PART TWO

Three Dimensions and Up

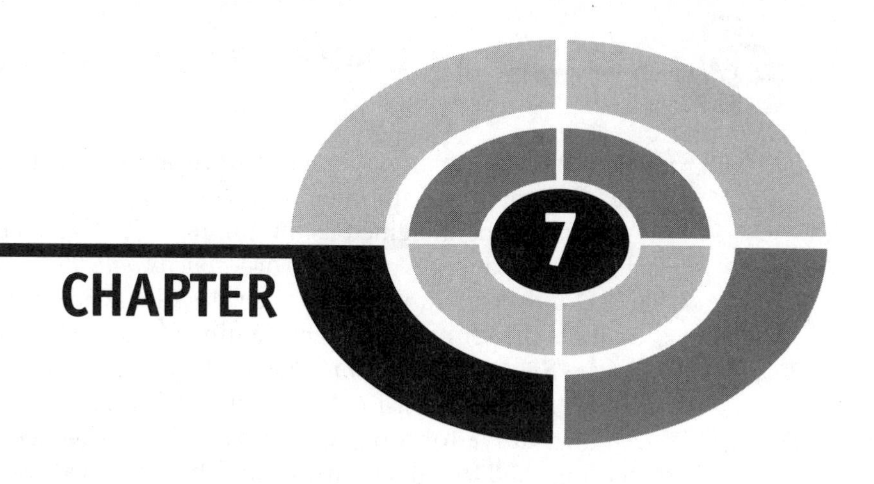

An Expanded Set of Rules

Solid geometry involves points, lines, and planes. The difference between plane geometry and solid geometry is the fact that there's an extra dimension. This means greater freedom, such as we would enjoy if we had flying cars. It also means that things are more complicated, reflecting the expanded range of maneuvers we would have to master if we had flying cars.

Points, Lines, Planes, and Space

A point in space can be envisioned as an infinitely tiny sphere, having height, width, and depth all equal to zero, but nevertheless possessing a specific location. A point is *zero-dimensional* (0D). A point in space is just like a point in a plane or a point on a line.

A line can be thought of as an infinitely thin, perfectly straight, infinitely long wire. It is *one-dimensional* (1D). A line in space is just like a line in a plane, but there are more possible directions in which lines can be oriented in space, as compared with lines confined to a plane.

A *plane* can be imagined as an infinitely thin, perfectly flat surface having an infinite expanse. A plane is *two-dimensional* (2D). A plane comprises a "flat 2D universe" in which all the rules of Euclidean plane geometry apply.

Space is the set of points representing all possible physical locations in the universe. Space is *three-dimensional* (3D). The idiosyncrasies of time, often called a "fourth dimension," are ignored in *Euclidean space*.

An alternative form of 3D can be defined in which there are two spatial dimensions and one time dimension. This type of three-space can be thought of as an Euclidean plane that has always existed, exists now, and always will exist.

If time is included in a concept of space, we get *four-dimensional* (4D) space, also known as *hyperspace*. We'll look at hyperspace later in this book. It, as you can imagine, gives us "hyperfreedom."

NAMING POINTS, LINES, AND PLANES

Points, lines, and planes in solid geometry are usually named using uppercase, italicized letters of the alphabet, just as they are in plane geometry. A common name for a point is P (for "point"). A common name for a line is L (for "line"). When it comes to planes in 3D space, we must use our imaginations. The letters X, Y, and Z are good choices. Sometimes lowercase, non-italic letters are used, such as m and n.

When we have two or more points, the letters immediately following P are used, for example Q, R, and S. If two or more lines exist, the letters immediately following L are used, for example M and N. Alternatively, numeric subscripts can be used. We might have points called P_1, P_2, P_3, and so forth, lines called L_1, L_2, L_3, and so forth, and planes called X_1, X_2, X_3, and so forth.

THREE POINT PRINCIPLE

Suppose that P, Q, and R are three different geometric points, no two of which lie on the same line. Then these points define one and only one (a unique or specific) plane X. The following two statements are always true, as shown in Fig. 7-1:

- P, Q, and R lie in a common plane X
- X is the only plane in which all three points lie

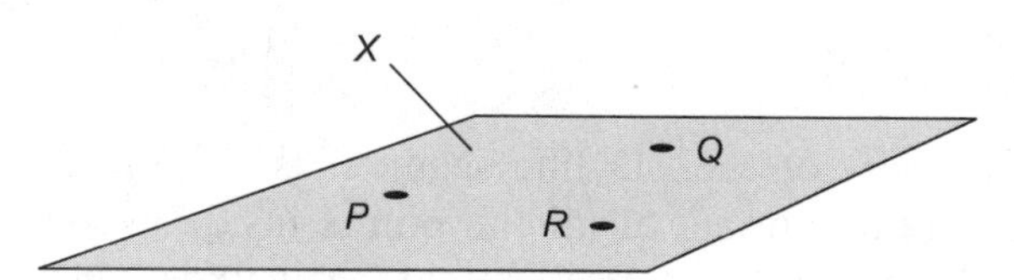

Fig. 7-1. Three points *P*, *Q*, and *R*, not all on the same line, define a specific plane *X*. The plane extends infinitely in 2D.

In order to show that a surface extends infinitely in 2D, we have to be imaginative. It's not as easy as showing that a line extends infinitely in 1D, because there aren't any good ways to draw arrows on the edges of a plane region the way we can draw them on the ends of a line segment. It is customary to draw planes as rectangles in perspective; they appear as rectangles, parallelograms, or trapezoids when rendered on a flat page. This is all right, as long as it is understood that the surface extends infinitely in 2D.

INTERSECTING LINE PRINCIPLE

Suppose that lines *L* and *M* intersect in a point *P*. Then the two lines define a unique plane *X*. The following two statements are always true, as shown in Fig. 7-2:

- *L* and *M* lie in a common plane *X*
- *X* is the only plane in which both lines lie

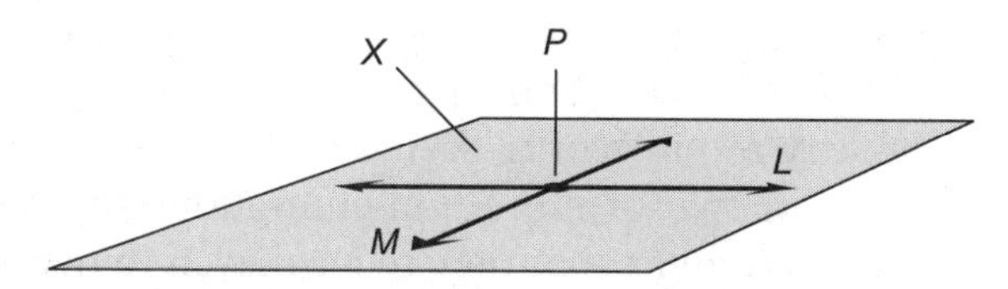

Fig. 7-2. Two lines *L* and *M*, intersecting at point *P*, define a specific plane *X*. The plane extends infinitely in 2D.

LINE AND POINT PRINCIPLE

Let *L* be a line and *P* be a point not on that line. Then line *L* and point *P* define a unique plane *X*. The following two statements are always true:

- *L* and *P* lie in a common plane *X*
- *X* is the only plane in which both *L* and *P* lie

PLANE REGIONS

The 2D counterpart of the 1D line segment is the *simple plane region*. A simple plane region consists of all the points inside a polygon or enclosed curve. The polygon or curve itself might be included in the set of points comprising the simple plane region, but this is not necessarily the case. If the polygon or curve is included, the simple plane region is said to be *closed*. Some examples are denoted in Fig. 7-3A; the boundaries are drawn so they look continuous. If the polygon or curve is not included, the simple plane region is said to be *open*. In Fig. 7-3B, several examples of open simple plane regions are denoted; the boundaries are dashed.

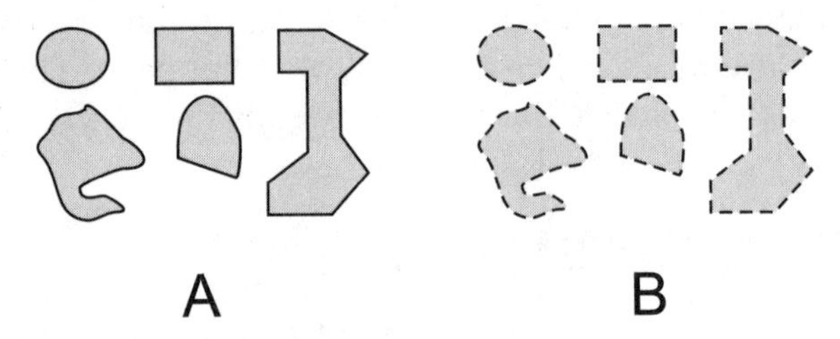

Fig. 7-3. Plane regions. At A, closed; at B, open.

The respective regions in Figs. 7-3A and 7-3B have identical shapes. They also have identical perimeters and identical interior areas. The outer boundaries do not add anything to the perimeter or the interior area.

These examples show specialized cases, in which the regions are contiguous, or "all of a piece," and the boundaries are either closed all the way around or open all the way around. Some plane regions have boundaries that are closed part of the way around, or in segments; it is also possible to have plane regions composed of two or more non-contiguous sub-regions. Some such plane regions are so complex that they're hard even to define. We won't concern ourselves with such monstrosities, other than to acknowledge that they can exist.

HALF PLANES

Sometimes, mathematicians talk about the portion of a geometric plane that lies "on one side" of a certain line. In Fig. 7-4, imagine the union of all possible geometric rays that start at L, then pass through line M (which is parallel to L), and extend onward past M forever in one direction. The region thus defined is known as a *half plane*.

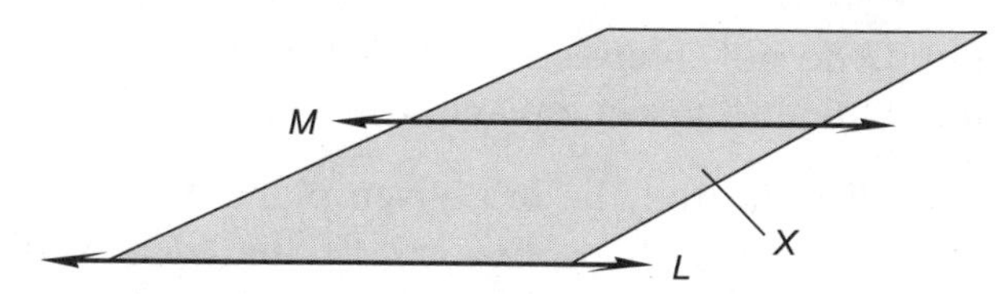

Fig. 7-4. A half plane X, defined by two parallel lines, L and M. The half plane extends infinitely in 2D on the "M" side of L.

The half plane defined by L and M might include the end line L, in which case it is *closed-ended.* In this case, line L is drawn as a solid line, as shown in Fig. 7-4. But the end line might not comprise part of the half plane, in which case the half plane is *open-ended.* Then line L is drawn as a dashed line.

Parts of the end line might be included in the half plane and other parts not included. There are infinitely many situations of this kind. Such scenarios are illustrated by having some parts of L appear solid, and other parts dashed.

INTERSECTING PLANES

Suppose that two different planes X and Y intersect; that is, they have points in common. Then the two planes intersect in a unique line L. The following two statements are always true, as shown in Fig. 7-5:

- Planes X and Y share a common line L
- L is the only line that planes X and Y have in common

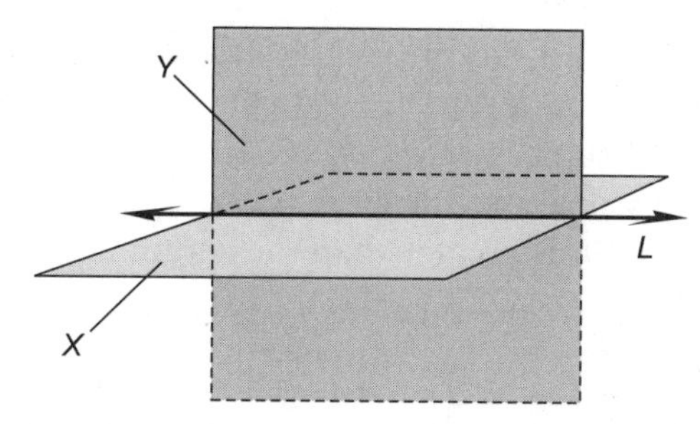

Fig. 7-5. The intersection of two planes X and Y determines a unique line L. The planes extend infinitely in 2D.

PARALLEL LINES IN 3D SPACE

By definition, two different lines L and M in three-space are *parallel lines* if and only if both of the following are true:

- Lines L and M do not intersect
- Lines L and M lie in the same plane X

If two lines are parallel and they lie in a given plane X, then X is the only plane in which the two lines lie. Thus, two parallel lines define a unique plane in Euclidean three-space.

SKEW LINES

By definition, two lines L and M in three-space are *skew lines* if and only if both of the following are true:

- Lines L and M do not intersect
- Lines L and M do not lie in the same plane

Imagine an infinitely long, straight, two-lane highway and an infinitely long, straight power line propped up on utility poles. Further imagine that the power line and the highway center line are both infinitely thin, and that the power line doesn't sag between the poles. Suppose the power line passes over the highway somewhere. Then the center line of the highway and the power line define skew lines.

PROBLEM 7-1

Find an example of a theoretical plane region with a finite, nonzero area but an infinite perimeter.

SOLUTION 7-1

Examine Fig. 7-6. Suppose the three lines PQ, RS, and TU (none of which are part of the plane region X, but are shown only for reference) are mutually parallel, and that the distances $d_1, d_2, d_3, \ldots$ are such that $d_2 = d_1/2$, $d_3 = d_2/2$, and in general, for any positive integer n, $d_{(n+1)} = d_n/2$. Also suppose that the length of line segment PV is greater than the length of line segment PT. Then

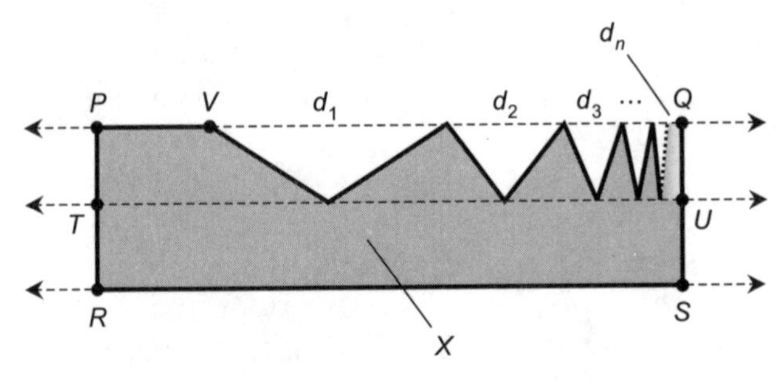

Fig. 7-6. Illustration for Problem 7-1.

plane region X has an infinite number of sides, each of which has a length greater than the length of line segment PT, so its perimeter is infinite. But the interior area of X is finite and nonzero, because it is obviously less than the interior area of quadrilateral $PQSR$ but greater than the area of quadrilateral $TUSR$.

PROBLEM 7-2

How many planes can mutually intersect in a given line L?

SOLUTION 7-2

In theory, an infinite number of planes can all intersect in a common line. Think of the line as an "Euclidean hinge," and then imagine a plane that can swing around this hinge. Each position of the "swinging plane" represents a unique plane in space.

Angles and Distances

Let's see how the *angles between intersecting planes* are defined, and how these angles behave. Let's also see how we can define the *angles between an intersecting line and plane*, and how these angles behave.

ANGLES BETWEEN INTERSECTING PLANES

Suppose two planes X and Y intersect in a common line L. Consider line M in plane X and line N in plane Y, such that $M \perp L$ and $N \perp L$, as shown in Fig. 7-7. The angle between the intersecting planes X and Y is called a *dihedral angle*, and can be represented in two ways. The first angle, whose measure is denoted by u, is the smaller angle between lines M and N. The second angle, whose measure is denoted by v, is the larger angle between lines M and N.

If only one angle is mentioned, the "angle between two intersecting planes" is usually considered to be the smaller angle u. Therefore, the angle of intersection is larger than zero but less than or equal to a right angle. That is, $0° < u \leq 90°$ ($0 < u \leq \pi/2$).

ADJACENT ANGLES BETWEEN INTERSECTING PLANES

Suppose two planes intersect, and their angles of intersection are u and v as defined above. Then if u and v are specified in degrees, $u + v = 180°$. If u and v are specified in radians, then $u + v = \pi$.

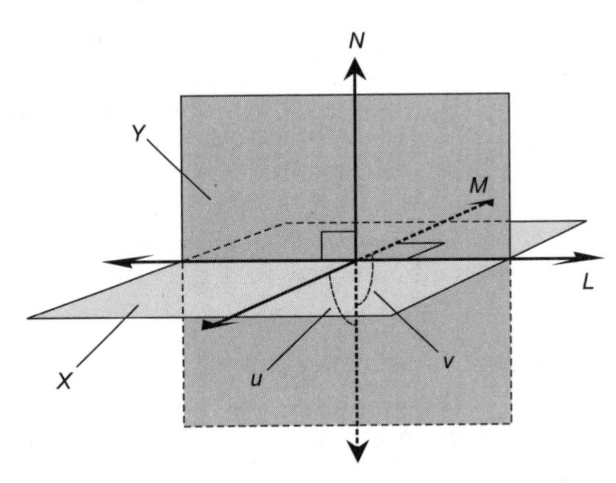

Fig. 7-7. The dihedral angle between the intersecting planes X and Y can be represented by u, the acute angle between lines M and N, or by v, the obtuse angle between lines M and N.

PERPENDICULAR PLANES

Suppose two planes X and Y intersect in a common line L. Consider line M in plane X and line N in plane Y, such that $M \perp L$ and $N \perp L$, as shown in Fig. 7-7. Then X and Y are said to be *perpendicular planes* if the angles between lines M and N are right angles, that is, $u = v = 90°$ ($\pi/2$ rad). Actually, it suffices to say that either $u = 90°$ ($\pi/2$ rad) or $v = 90°$ ($\pi/2$ rad).

NORMAL LINE TO A PLANE

Let plane X be determined by lines L and M, which intersect at point S. Then line N that passes through plane X at point S is *normal* (also called *perpendicular* or *orthogonal*) to plane X if and only if $N \perp L$ and $N \perp M$. This is shown in Fig. 7-8. Line N is the only line normal to plane X at point S. Furthermore, line N is perpendicular to any line, line segment, or ray that lies in plane X and runs through point S.

ANGLES BETWEEN AN INTERSECTING LINE AND PLANE

Let X be a plane. Suppose a line O, which is not normal to plane X, intersects plane X at some point S as shown in Fig. 7-9. In order to define an angle at which line O intersects plane X, we must construct some objects. Let N be a line normal to plane X, passing through point S. Let Y be the plane deter-

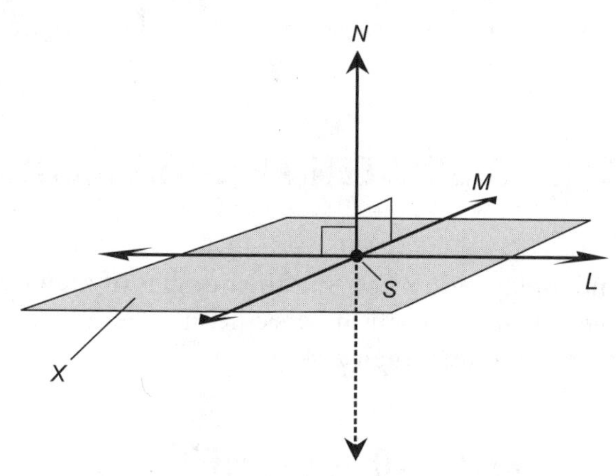

Fig. 7-8. Line N through plane X at point S is normal to X if and only if $N \perp L$ and $N \perp M$, where L and M are lines in plane X that intersect at point S.

mined by the intersecting lines N and O. Let L be the line formed by the intersection of planes X and Y. The angle between line O and plane X can be represented in two ways. The first angle, whose measure is denoted by u, is the smaller angle between lines L and O as determined in plane Y. The second angle, whose measure is denoted by v, is the larger angle between lines L and O as determined in plane Y.

If only one angle is mentioned, the "angle between a line and a plane that intersect" is considered to be the smaller angle u. Therefore, the angle of

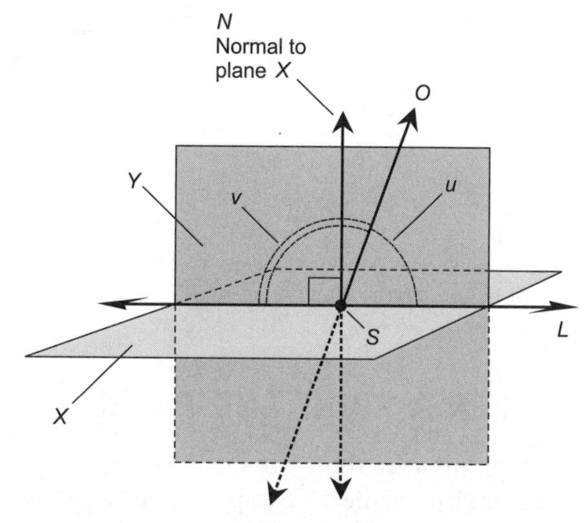

Fig. 7-9. Angles u and v between a plane X and a line O that passes through X at point S.

intersection is larger than zero but less than or equal to a right angle. That is, $0° < u \leq 90°$ ($0 < u \leq \pi/2$).

ADJACENT ANGLES BETWEEN AN INTERSECTING LINE AND PLANE

Suppose a line and a plane intersect, and their angles of intersection are u and v as defined above. Then if u and v are specified in degrees, $u + v = 180°$. If u and v are specified in radians, then $u + v = \pi$.

DROPPING A NORMAL TO A PLANE

Suppose that R is a point near, but not in, a plane X. Then there is exactly one line N through point R, intersecting plane X at some point S, such that line N is normal to plane X, as shown in Fig. 7-10. Any lines in plane X that pass through point S, such as L and M shown in the figure, must necessarily be perpendicular to line N.

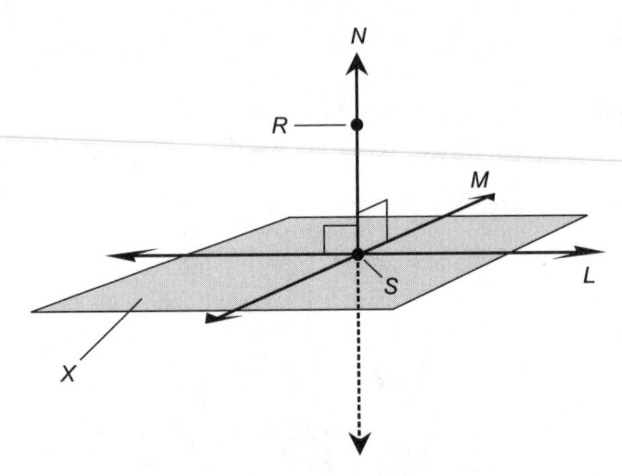

Fig. 7-10. Line N through point R is normal to plane X at point S. The distance between R and X is equal to the length of line segment RS.

DISTANCE BETWEEN A POINT AND PLANE

Suppose that R is a point near, but not in, a plane X. Let N be the line through R that is normal to plane X. Suppose line N intersects plane X at point S. Then the distance between point R and plane X is equal to the length of line segment RS (Fig. 7-10).

PLANE PERPENDICULAR TO LINE

Imagine a line N in space. Imagine a specific point S on line N. There is exactly one plane X containing point S, such that line N is normal to plane X at point S (Fig. 7-10). Any lines in plane X that pass through point S, such as L and M shown in the figure, must necessarily be perpendicular to line N.

LINE PARALLEL TO PLANE

A line L is parallel to a plane X if and only if the following two conditions hold true:

- Line L does not lie in plane X
- Line L does not intersect plane X

Under these conditions, there is exactly one line M in plane X, such that lines L and M are parallel. Any line N in plane X, other than line M, is a skew line relative to L (Fig. 7-11).

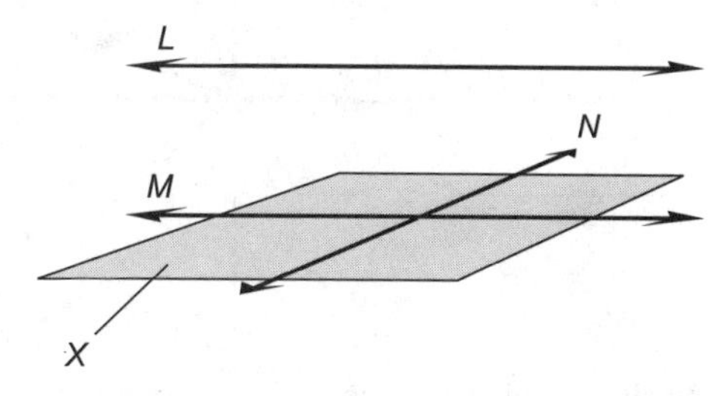

Fig. 7-11. A line L parallel to a plane X. There is exactly one line M in plane X such that M is parallel to L; all other lines N in plane X are skew lines relative to L.

DISTANCE BETWEEN A PARALLEL LINE AND PLANE

Suppose line L is parallel to plane X. Let R be some (any) point on line L. Then the distance between line L and plane X is equal to the distance between point R and plane X, which has already been defined.

ADDITION AND SUBTRACTION OF ANGLES BETWEEN INTERSECTING PLANES

Angles between intersecting planes add and subtract just like angles between intersecting lines (or line segments). Here is how we can prove it, based on knowledge we already have.

Suppose three planes X, Y, and Z intersect in a single, common line L. Let S be a point on line L. Let P, Q, and R be points on planes X, Y, and Z respectively, such that line segments SP, SQ, and SR are all perpendicular to line L. Let $\angle XY$ be the angle between planes X and Y, $\angle YZ$ be the angle between planes Y and Z, and $\angle XZ$ be the angle between planes X and Z. This is diagrammed in Fig. 7-12. From the preceding definition of the angle between two planes, we know that:

$$\angle XY = \angle PSQ$$
$$\angle YZ = \angle QSR$$
$$\angle XZ = \angle PSR$$

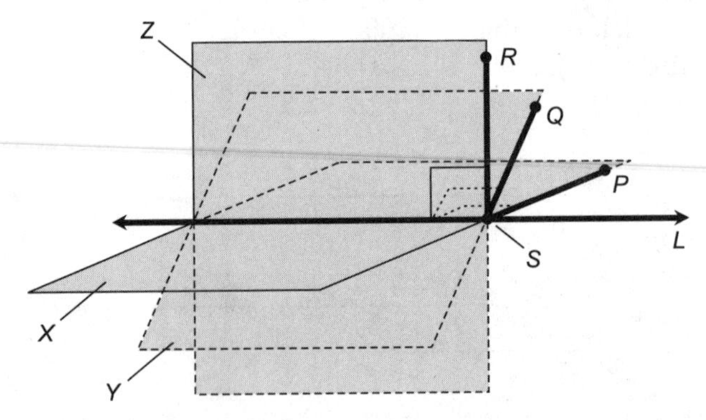

Fig. 7-12. Addition and subtraction of angles between planes.

Line segments SP, SQ, and SR all lie in a single plane because they all intersect at point S and they are all perpendicular to line L. Therefore, we know from the rules for addition of angles in a plane, that the following hold true for the measures of the angles between the line segments:

$$\angle PSQ + \angle QSR = \angle PSR$$
$$\angle PSR - \angle QSR = \angle PSQ$$
$$\angle PSR - \angle PSQ = \angle QSR$$

Substituting the angles between the planes for the angles between the line segments, we see that the following hold true for the measures of the angles between the planes:

$$\angle XY + \angle YZ = \angle XZ$$
$$\angle XZ - \angle YZ = \angle XY$$
$$\angle XZ - \angle XY = \angle YZ$$

PROBLEM 7-3

Suppose a communications cable is strung above a fresh-water lake. Imagine that the cable does not sag at all, and is attached at the tops of a set of utility poles. Suppose the engineering literature recommends that the cable be suspended 10 meters above "effective ground," and that "effective ground" is, on average, 2 meters below the average level of the surface of a body of fresh water. How tall should the poles be? Assume they are all perfectly vertical.

SOLUTION 7-3

Because the poles are perfectly vertical, they are perpendicular to the surface of the lake. This means that the poles should each be tall enough so their tops are 10 meters above "effective ground," so they should each extend 10 – 2 meters, or 8 meters, above the water surface. The actual height of each pole depends on the depth of the lake at the point where it is placed. It is assumed that the lake is small enough, and/or weather conditions reasonable enough, so the lake does not acquire waves so high that they inundate the cable!

PROBLEM 7-4

Imagine that you are flying a kite over a perfectly flat field. The kite is of a design that flies at a "high angle." Suppose the kite line does not sag, and the kite flies only 10° away from the vertical. (Some kites can actually fly straight overhead!) Imagine that it is a sunny day, and the sun is shining down from exactly the zenith. What is the angle between the kite string (also called the kite line) and its shadow on the flat field?

SOLUTION 7-4

Suppose you stand at point S on the surface of the field, which we call plane X. The kite line and its shadow lie along lines SR and ST, as shown in Fig. 7-13. The sun shines down along a line QS that is normal to plane X. Lines SQ, SR, and ST all lie in a common plane Y, which is perpendicular to plane X. We know that the measure of $\angle QSR$ is 10°, because we are given this information. We also know that the measure of $\angle QST$ is 90°, because line QS is normal to plane X, and line ST lies in plane X. Because lines SR, ST, and SQ

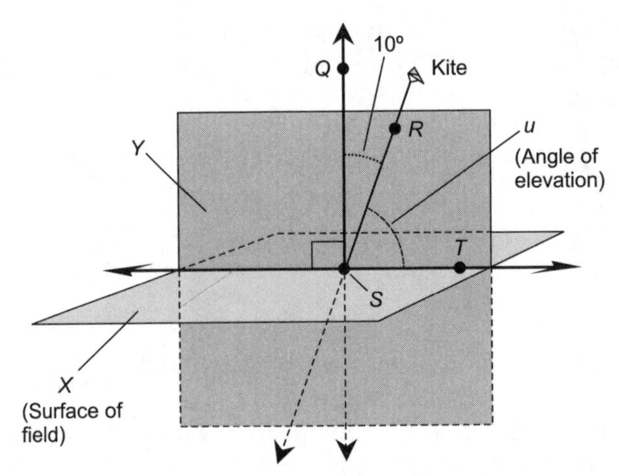

Fig. 7-13. Illustration for Problem 7-4.

all lie in the same plane *Y*, we know that the measures of the angles among them add as follows:

$$\angle QSR + \angle RST = \angle QST$$

and therefore

$$\angle RST = \angle QST - \angle QSR$$

The measure of $\angle RST$, which represents the angle between the kite line and its shadow, is equal to 90° – 10°, or 80°.

More Facts

In the confines of a single geometric plane, lines and angles behave according to various rules. The following are some of the best-known examples.

PARALLEL PLANES

Two distinct planes are *parallel* if and only if they do not intersect. Two distinct half planes are parallel if and only if the planes in which they lie do not intersect. Two distinct plane regions are parallel if and only if the planes in which they lie do not intersect.

DISTANCE BETWEEN PARALLEL PLANES

Suppose planes X and Y are parallel. Let R be some arbitrary point on plane X. Then the distance between planes X and Y is equal to the distance between point R and plane Y, which has already been defined.

VERTICAL ANGLES FOR INTERSECTING PLANES

Suppose that Y and Z are two planes that intersect in a line L. Let points P, Q, R, S, and T be as shown in Fig. 7-14, such that:

- Point T lies on lines L, PS, and RQ
- Points Q and R lie in plane Y
- Points P and S lie in plane Z
- Lines PS and RQ are both perpendicular to line L

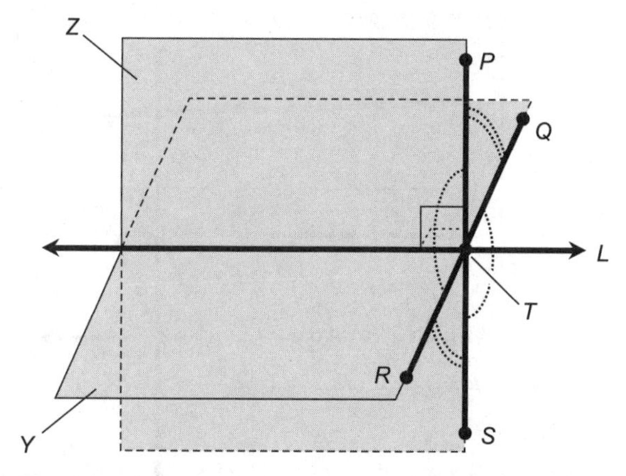

Fig. 7-14. Vertical angles between intersecting planes.

Then the pair $\angle PTQ$ and $\angle STR$ are vertical angles; the pair $\angle RTP$ and $\angle QTS$ are also vertical angles. Thus, $\angle PTQ$ has the same measure as $\angle STR$, and $\angle RTP$ has the same measure as $\angle QTS$.

ALTERNATE INTERIOR ANGLES FOR INTERSECTING PLANES

Suppose that X is a plane that passes through two parallel planes Y and Z, intersecting Y and Z in lines L and M. Let points P, Q, R, S, T, U, V, and W be as shown in Figs. 7-15A and 7-15B, such that:

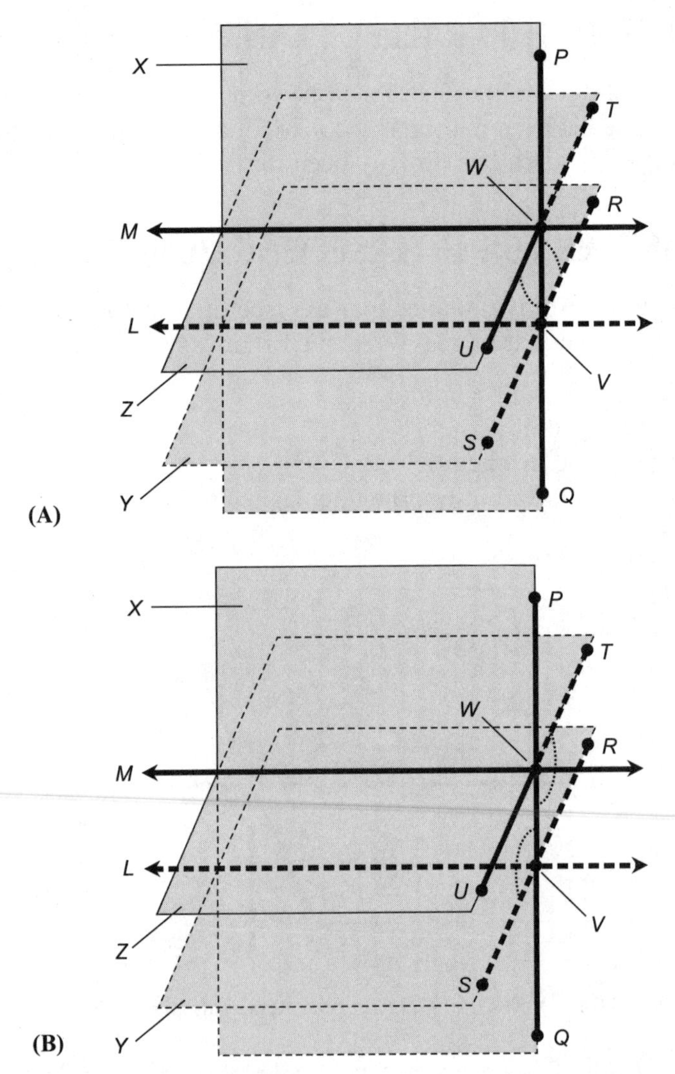

Fig. 7-15. (A) Alternate interior angles between intersecting planes. (B) Another example of alternate interior angles between intersecting planes.

- Point V lies on lines L, PQ, and RS
- Point W lies on lines M, PQ, and TU
- Points P and Q lie in plane X
- Points R and S lie in plane Y
- Points T and U lie in plane Z
- Lines PQ and RS are perpendicular to line L
- Lines PQ and TU are perpendicular to line M

Then the pair $\angle PVR$ and $\angle QWU$ are alternate interior angles (Fig. 7-15A); the pair $\angle TWQ$ and $\angle SVP$ are also alternate interior angles (Fig. 7-15B). Alternate interior angles always have equal measures. Thus, $\angle PVR$ has the same measure as $\angle QWU$, and $\angle TWQ$ has the same measure as $\angle SVP$.

ALTERNATE EXTERIOR ANGLES FOR INTERSECTING PLANES

Suppose that X is a plane that passes through two parallel planes Y and Z, intersecting Y and Z in lines L and M. Let points P, Q, R, S, T, U, V, and W be as shown in Fig. 7-16, such that:

- Point V lies on lines L, PQ, and RS
- Point W lies on lines M, PQ, and TU
- Points P and Q lie in plane X
- Points R and S lie in plane Y
- Points T and U lie in plane Z
- Lines PQ and RS are perpendicular to line L
- Lines PQ and TU are perpendicular to line M

Then the pair $\angle PWT$ and $\angle QVS$ are alternate exterior angles; the pair $\angle UWP$ and $\angle RVQ$ are also alternate exterior angles. Alternate exterior angles always have equal measures. Thus, $\angle PWT$ has the same measure as $\angle QVS$, and $\angle UWP$ has the same measure as $\angle RVQ$.

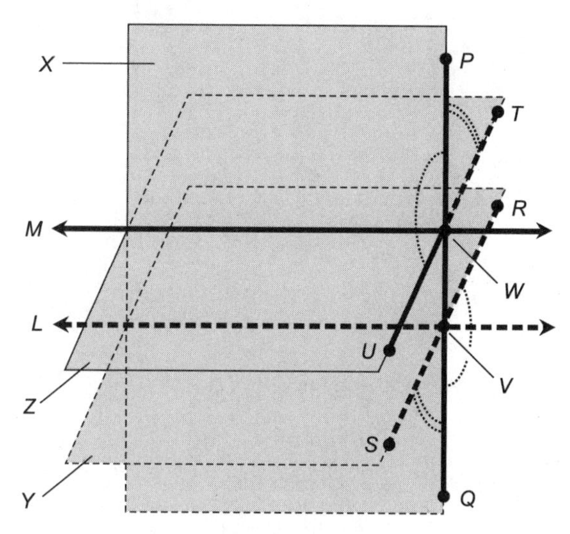

Fig. 7-16. Alternate exterior angles between intersecting planes.

CORRESPONDING ANGLES FOR INTERSECTING PLANES

Suppose that X is a plane that passes through two parallel planes Y and Z, intersecting Y and Z in lines L and M. Let points P, Q, R, S, T, U, V, and W be as shown in Fig. 7-17, such that:

- Point V lies on lines L, PQ, and RS
- Point W lies on lines M, PQ, and TU
- Points P and Q lie in plane X
- Points R and S lie in plane Y
- Points T and U lie in plane Z
- Lines PQ and RS are perpendicular to line L
- Lines PQ and TU are perpendicular to line M

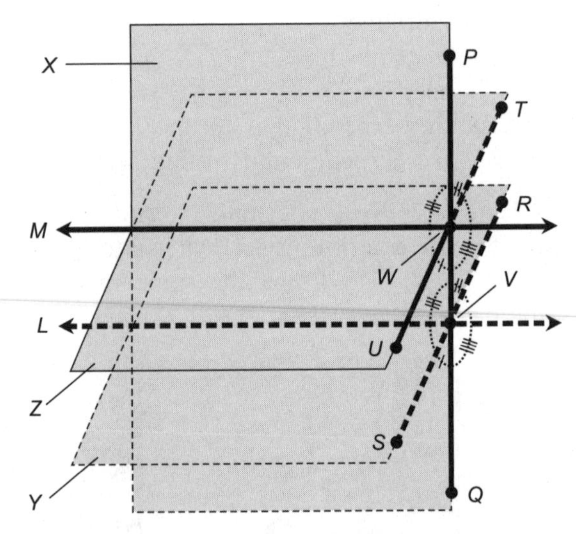

Fig. 7-17. Corresponding angles between intersecting planes.

Then the following pairs of angles are corresponding angles, and each pair has equal measures:

$$\angle QWU = \angle QVS$$
$$\angle PWT = \angle PVR$$
$$\angle UWP = \angle SVP$$
$$\angle TWQ = \angle RVQ$$

PARALLEL PRINCIPLE FOR PLANES

Suppose X is a plane and R is a point not on X. Then there exists one and only one plane Y through R such that plane Y is parallel to plane X. This is the 3D counterpart of the parallel principle for Euclidean plane geometry.

The denial of this principle, which is in fact a postulate, can take two forms:

- There exist no planes Y through point R such that plane Y is parallel to plane X
- There exist infinitely many planes Y through point R such that plane Y is parallel to plane X

Both of these hypotheses give rise to forms of non-Euclidean geometry in which 3D space is "curved" or "warped." Albert Einstein was one of the first scientists to envision a universe in which space is non-Euclidean.

PARALLEL PRINCIPLE FOR LINES AND PLANES

Suppose X is a plane and R is a point not on X. Then there exist an infinite number of lines through R that are parallel to plane X. All of these lines lie in the plane Y through R such that plane Y is parallel to plane X.

The denial of the parallel principle for planes, defined in the previous paragraph, can result in there being no lines through R that are parallel to plane X. In certain specialized instances, it can even result in there being exactly one line through R that is parallel to plane X. If you have trouble imagining this, don't be concerned. You must think in 4D, and that is a mental trick that few humans can do until they become armed with the power of non-Euclidean mathematics.

PROBLEM 7-5

Suppose you are standing inside a large warehouse. The floor is flat and level, and the ceiling is flat and is at a uniform height of 5.455 meters above the floor. You have a flashlight with a narrow beam, and hold it so its bulb is 1.025 meters above the floor. You shine the beam upward toward the ceiling. The center of the beam strikes the ceiling 9.577 meters from the point on the ceiling directly above the bulb. How long is the shaft of light representing the center of the beam? Round your answer off to two decimal places.

SOLUTION 7-5

Let's call the flashlight bulb point A, the point at which the center of the light beam strikes the ceiling point B, and the point directly over the flashlight bulb point C as shown in Fig. 7-18. Call the lengths of the sides opposite these points a, b, and c. Then $\triangle ABC$ is a right triangle because line segment AC (whose length is b) is normal to the ceiling at point C, and therefore is perpendicular to line segment BC. The right angle is $\angle ACB$. From this, we know that the lengths of the sides are related according to the Pythagorean formula:

$$a^2 + b^2 = c^2$$

We want to know the length of side c; therefore:

$$c = (a^2 + b^2)^{1/2}$$

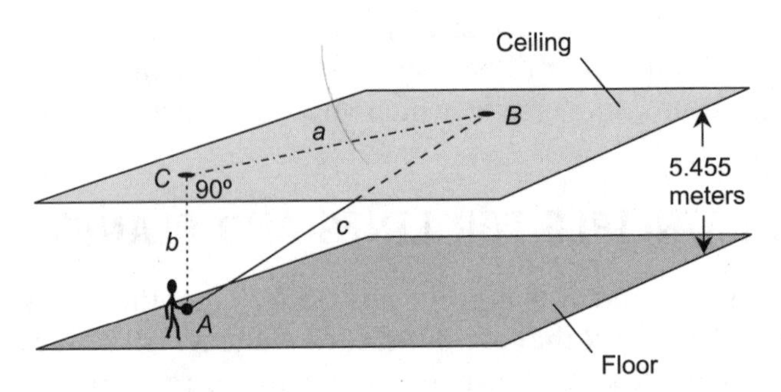

Fig. 7-18. Illustration for Problem 7-5.

Calling meters "units" so we don't have to write the word "meters" over and over, the length of side a is given as 9.577. The length of side b is equal to the height of the ceiling above the floor, minus the height of the bulb above the floor:

$$b = 5.455 - 1.025 = 4.430$$

Therefore:

$$\begin{aligned} c &= (9.577^2 + 4.430^2)^{1/2} \\ &= (91.719 + 19.625)^{1/2} \\ &= 111.344^{1/2} \\ &= 10.55 \end{aligned}$$

The center of the beam is 10.55 meters long.

Quiz

Refer to the text in this chapter if necessary. A good score is eight correct. Answers are in the back of the book.

1. Which of the following statements is false?
   (a) If a line is normal to one of two parallel planes, and that line does not lie in either plane, then that line is normal to the other of the two parallel planes
   (b) If a line is parallel to one of two parallel planes, and that line does not lie in either plane, then that line is parallel to the other of the two parallel planes
   (c) In 3D space, there cannot exist three different planes, each of which is perpendicular to the other two
   (d) In 3D space, there can exist infinitely many different planes, each of which is parallel to all the others

2. Suppose three different planes X_1, X_2, and X_3 intersect in a single line L. Imagine a fourth plane X_4 that is perpendicular to all three of the planes X_1, X_2, and X_3. Which of the following statements is true?
   (a) Line L is normal to plane X_4
   (b) Line L is parallel to plane X_4
   (c) Any line in plane X_4 is a skew line relative to L
   (d) There can exist no plane X_4 with the characteristics described

3. Suppose that in a certain 3D universe called U, it is impossible to find a plane Y that is parallel to a certain plane X. From this, we can conclude that
   (a) all planes in U are flat
   (b) half of the planes in U are curved
   (c) U is non-Euclidean
   (d) time is warped in U

4. Suppose two different planes X_1 and X_2 intersect in a single line L. Consider a line M that is normal to both planes X_1 and X_2. Which of the following statements is true?
   (a) Line M must be normal to either X_1 or X_2
   (b) Line M must lie in either X_1 or X_2
   (c) Line M must intersect line L
   (d) There can exist no line M with the characteristics described

5. Imagine that you are in a huge warehouse in which the floor and ceiling are both level and flat. You place a flashlight on the floor so the center of its beam shines upward at a 20° angle relative to the floor. The beam strikes the ceiling some distance away. At what angle, with respect to a normal line to the ceiling, does the center of the beam strike the ceiling?
   (a) 20°
   (b) 70°
   (c) 160°
   (d) It is impossible to tell without more information

6. Imagine you are in a huge warehouse in which the floor and ceiling are both level and flat, the ceiling is made of glass, and the ceiling forms the roof of the warehouse. You place a flashlight on the floor so the center of its beam shines upward at a 40° angle relative to the floor. The beam strikes and passes through the glass roof some distance away. At what angle, with respect to the plane of the roof, will the center of the beam emerge from the glass?
   (a) 40°
   (b) 50°
   (c) 130°
   (d) It is impossible to tell without more information

7. Suppose a closed plane region, having the shape of an octagon, has interior area equal to k square units. Suppose the perimeter of the region is m units. The interior area of the open region (the octagon not including the outer boundary) is
   (a) k square units
   (b) $k - m$ square units
   (c) $k - m$ units
   (d) impossible to determine without more information

8. A plane can be uniquely defined by
   (a) three points
   (b) two intersecting lines
   (c) a line and a point not on that line
   (d) any of the above

9. Suppose two different planes X_1 and X_2 intersect in a single line L. Consider a line M that is parallel to both planes X_1 and X_2. Which of the following statements is true?
   (a) Line M must lie in either X_1 or X_2
   (b) Line M must lie in both X_1 and X_2

(c) Line M must lie outside both X_1 and X_2
(d) There can exist no line M with the characteristics described

10. Suppose two planes intersect at an angle of 70°. This is the more common of two ways the intersection angle can be expressed. The less common value for the intersection angle of these two planes is
(a) 20°
(b) 180°
(c) 290°
(d) none of the above

Surface Area and Volume

In this chapter, you'll learn how to find the surface areas and volumes of various geometric solids in Euclidean three-space. Only the simplest sorts of objects are dealt with here.

Straight-Edged Objects

In Euclidean three-space, geometric solids with straight edges have flat faces, also called *facets*, each of which forms a plane polygon. An object of this sort is known as a *polyhedron*.

THE TETRAHEDRON

A polyhedron in 3D must have at least four faces. A four-faced polyhedron is called a *tetrahedron*. Each of the four faces of a tetrahedron is a triangle.

There are four vertices. Any four specific points, not all in a single plane, form a unique tetrahedron.

SURFACE AREA OF TETRAHEDRON

Figure 8-1 shows a tetrahedron. The surface area is found by adding up the interior areas of all four triangular faces. In the case of a *regular tetrahedron*, all six edges have the same length, and therefore all four faces are equilateral triangles. If the length of each edge of a regular tetrahedron is equal to s units, then the surface area, B, of the whole four-faced regular tetrahedron is given by:

$$B = 3^{1/2} s^2$$

where $3^{1/2}$ represents the square root of 3, or approximately 1.732. This also happens to be twice the sine of 60°, which is the angle between any two edges of the figure.

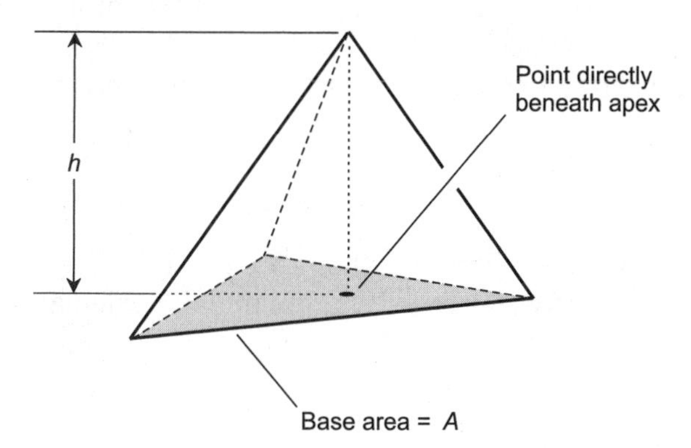

Fig. 8-1. A tetrahedron has four faces (including the base) and six edges.

VOLUME OF TETRAHEDRON

Imagine a tetrahedron whose base is a triangle with area A, and whose height is h as shown in Fig. 8-1. The volume, V, of the figure is given by:

$$V = Ah/3$$

THE PYRAMID

Figure 8-2 illustrates a *pyramid*. This figure has a square or rectangular base and four slanted faces. If the base is a square and the apex (the top of the

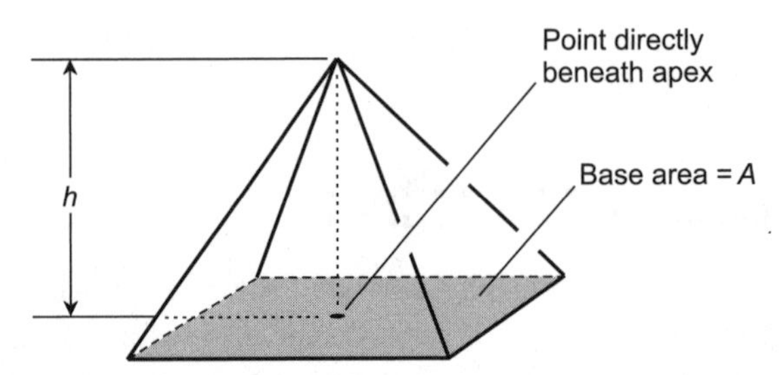

Fig. 8-2. A pyramid has five faces (including the base) and eight edges.

pyramid) lies directly above a point at the center of the base, then the figure is a *regular pyramid*, and all of the slanted faces are isosceles triangles.

SURFACE AREA OF PYRAMID

The surface area of a pyramid is found by adding up the areas of all five of its faces (the four slanted faces plus the base). In the case of a regular pyramid where the length of each slanted edge, called the *slant height*, is equal to s units and the length of each edge of the base is equal to t units, the surface area, B, is given by:

$$B = t^2 + 2t\,(s^2 - t^2/4)^{1/2}$$

In the case of an *irregular pyramid*, the problem of finding the surface area is more complicated, because it involves individually calculating the area of the base and each slanted face, and then adding all the areas up.

VOLUME OF PYRAMID

Imagine a pyramid whose base is a square with area A, and whose height is h as shown in Fig. 8-2. The volume, V, of the pyramid is given by:

$$V = Ah/3$$

This holds true whether the pyramid is regular or irregular.

THE CUBE

Figure 8-3 illustrates a *cube*. This is a *regular hexahedron* (six-sided polyhedron). It has 12 edges, each of which is of the same length. Each of the six faces is a square.

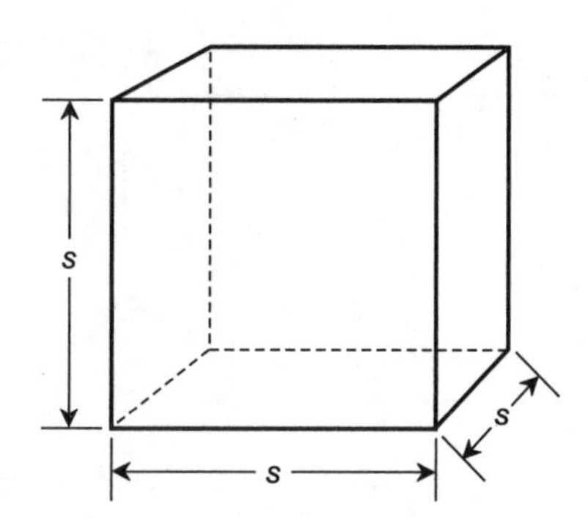

Fig. 8-3. A cube has six square faces and 12 edges of identical length.

SURFACE AREA OF CUBE

Imagine a cube whose edges each have length s, as shown in Fig. 8-3. The surface area, A, of the cube is given by:

$$A = 6s^2$$

VOLUME OF CUBE

Imagine a cube as defined above and in Fig. 8-3. The volume, V, of the solid enclosed by the cube is given by:

$$V = s^3$$

THE RECTANGULAR PRISM

Figure 8-4 illustrates a *rectangular prism*. This is a hexahedron, each of whose six faces is a rectangle. The figure has 12 edges, but they are not necessarily all equally long.

SURFACE AREA OF RECTANGULAR PRISM

Imagine a rectangular prism whose edges have lengths s_1, s_2, and s_3 as shown in Fig. 8-4. The surface area, A, of the prism is given by:

$$A = 2s_1s_2 + 2s_1s_3 + 2s_2s_3$$

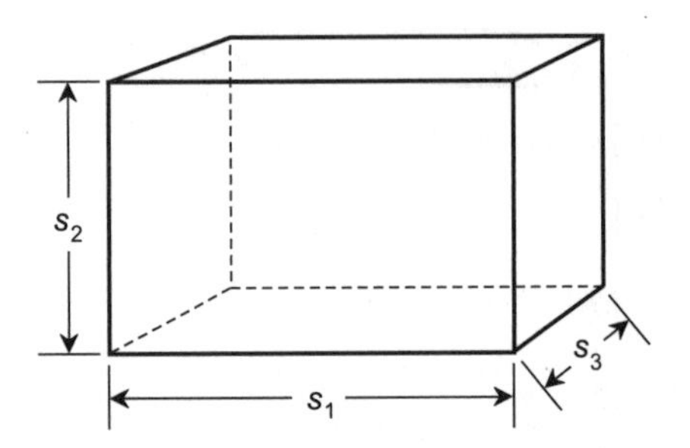

Fig. 8-4. A rectangular prism has six rectangular faces and 12 edges.

VOLUME OF RECTANGULAR PRISM

Imagine a rectangular prism as defined above and in Fig. 8-4. The volume, V, of the enclosed solid is given by:

$$V = s_1 s_2 s_3$$

THE PARALLELEPIPED

A *parallelepiped* is a six-faced polyhedron in which each face is a parallelogram, and opposite pairs of faces are congruent. The figure has 12 edges. The acute angles between the pairs of edges are x, y, and z, as shown in Fig. 8-5.

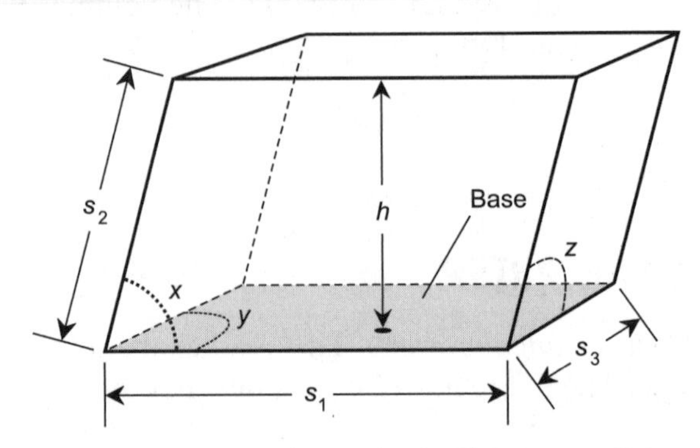

Fig. 8-5. A parallelepiped has six faces, all of which are parallelograms, and 12 edges.

SURFACE AREA OF PARALLELEPIPED

Imagine a parallelepiped with edges of lengths s_1, s_2, and s_3. Suppose the angles between pairs of edges are x, y, and z as shown in Fig. 8-5. The surface area, A, of the parallelepiped is given by:

$$A = 2s_1s_2 \sin x + 2s_1s_3 \sin y + 2s_2s_3 \sin z$$

where sin x represents the sine of angle x, sin y represents the sine of angle y, and sin z represents the sine of angle z.

VOLUME OF PARALLELEPIPED

Imagine a parallelepiped whose edges have lengths s_1, s_2, and s_3, and that has angles between edges of x, y, and z as shown in Fig. 8-5. Suppose further that the height of the parallelepiped, as measured along a line normal to the base, is equal to h. The volume, V, of the enclosed solid is equal to the product of the base area and the height:

$$V = hs_1s_3 \sin y$$

PROBLEM 8-1

Suppose you want to paint the interior walls of a room in a house. The room is shaped like a rectangular prism. The ceiling is exactly 3.0 meters above the floor. The floor and the ceiling both measure exactly 4.2 meters by 5.5 meters. There are two windows, the outer frames of which both measure 1.5 meters high by 1.0 meter wide. There is one doorway, the outer frame of which measures 2.5 meters high by 1.0 meter wide. With two coats of paint (which you intend to apply), one liter of paint can be expected to cover exactly 20 square meters of wall area. How much paint, in liters, will you need to completely do the job?

SOLUTION 8-1

It is necessary to find the amount of wall area that this room has. Based on the information given, we can conclude that the rectangular prism formed by the edges between walls, floor, and ceiling measures 3.0 meters high by 4.2 meters wide by 5.5 meters deep. So we can let $s_1 = 3.0$, $s_2 = 4.2$, and $s_3 = 5.5$ (with all units assumed to be in meters) to find the surface area A of the rectangular prism, in square meters, neglecting the area subtracted by the windows and doorway. Using the formula:

$$\begin{aligned}
A &= 2s_1s_2 + 2s_1s_3 + 2s_2s_3 \\
&= (2 \times 3.0 \times 4.2) + (2 \times 3.0 \times 5.5) + (2 \times 4.2 \times 5.5) \\
&= 25.2 + 33.0 + 46.2 \\
&= 104.4 \text{ square meters}
\end{aligned}$$

There are two windows measuring 1.5 meters by 1.0 meter; each of these therefore takes away $1.5 \times 1.0 = 1.5$ square meters of area. The doorway measures 2.5 meters by 1.0 meter, so it takes away $2.5 \times 1.0 = 2.5$ square meters. Thus the windows and doorway combined take away $1.5 + 1.5 + 2.5 = 5.5$ square meters of wall space. Then we must also take away the areas of the floor and ceiling. This is the final factor in the above equation, $2s_2s_3 = 46.2$. The wall area to be painted, call it A_w, is therefore:

$$A_w = (104.4 - 5.5) - 46.2$$
$$= 52.7 \text{ square meters}$$

A liter of paint can be expected to cover 20 square meters. Therefore, we will need 52.7/20, or 2.635, liters of paint to do this job.

Cones and Cylinders

A *cone* has a circular or elliptical base and an apex point. The cone itself consists of the union of the following sets of points:

- The circle or ellipse
- All points inside the circle or ellipse and that lie in its plane
- All line segments connecting the circle or ellipse (not including its interior) and the apex point

The interior of the cone consists of the set of all points within the cone. The cone itself might or might not be included in the definition of the interior.

A *cylinder* has a circular or elliptical base, and a circular or elliptical top that is congruent to the base and that lies in a plane parallel to the base. The cylinder itself consists of the union of the following sets of points:

- The base circle or ellipse
- All points inside the base circle or ellipse and that lie in its plane
- The top circle or ellipse
- All points inside the top circle or ellipse and that lie in its plane
- All line segments connecting corresponding points on the base circle or ellipse and top circle or ellipse (not including their interiors)

The interior of the cylinder consists of the set of all points within the cylinder. The cylinder itself might or might not be included in the definition of the interior.

These are general definitions, and they encompass a great variety of objects! In this chapter, we'll look only at cones and cylinders whose bases are circles.

THE RIGHT CIRCULAR CONE

A *right circular cone* has a base that is a circle, and an apex point that lies on a line normal to the plane of the base and that passes through the center of the base (Fig. 8-6).

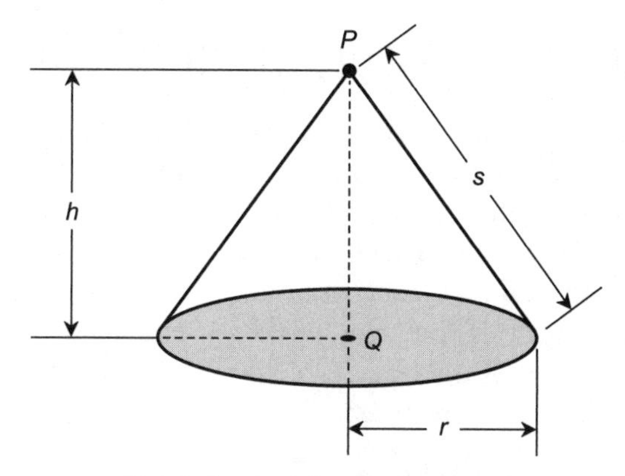

Fig. 8-6. A right circular cone.

SURFACE AREA OF RIGHT CIRCULAR CONE

Imagine a right circular cone as shown in Fig. 8-6. Let P be the apex of the cone, and let Q be the center of the base. Let r be the radius of the base, let h be the height of the cone (the length of line segment PQ), and let s be the slant height of the cone as measured from any point on the edge of the base to the apex P. The surface area S_1 of the cone, including the base, is given by either of the following formulas:

$$S_1 = \pi r^2 + \pi rs$$
$$S_1 = \pi r^2 + \pi r(r^2 + h^2)^{1/2}$$

The surface area S_2 of the cone, not including the base, is called the *lateral surface area* and is given by either of the following:

$$S_2 = \pi rs$$
$$S_2 = \pi r(r^2 + h^2)^{1/2}$$

VOLUME OF RIGHT CIRCULAR CONE

Imagine a right circular cone as defined above and in Fig. 8-6. The volume, V, of the interior of the figure is given by:

$$V = \pi r^2 h/3$$

SURFACE AREA OF FRUSTUM OF RIGHT CIRCULAR CONE

Imagine a right circular cone that is truncated (cut off) by a plane parallel to the base. This is called a *frustum* of the right circular cone. Let P be the center of the circle defined by the truncation, and let Q be the center of the base, as shown in Fig. 8-7. Suppose line segment PQ is perpendicular to the base. Let r_1 be the radius of the top, let r_2 be the radius of the base, let h be the height of the object (the length of line segment PQ), and let s be the slant height. Then the surface area S_1 of the object (including the base and the top) is given by either of the following formulas:

$$S_1 = \pi(r_1 + r_2)[h^2 + (r_2 - r_1)^2]^{1/2} + \pi(r_1^2 + r_2^2)$$
$$S_1 = \pi s(r_1 + r_2) + \pi(r_1^2 + r_2^2)$$

The surface area S_2 of the object (not including the base or the top) is given by either of the following:

$$S_2 = \pi(r_1 + r_2)[h^2 + (r_2 - r_1)^2]^{1/2}$$
$$S_2 = \pi s(r_1 + r_2)$$

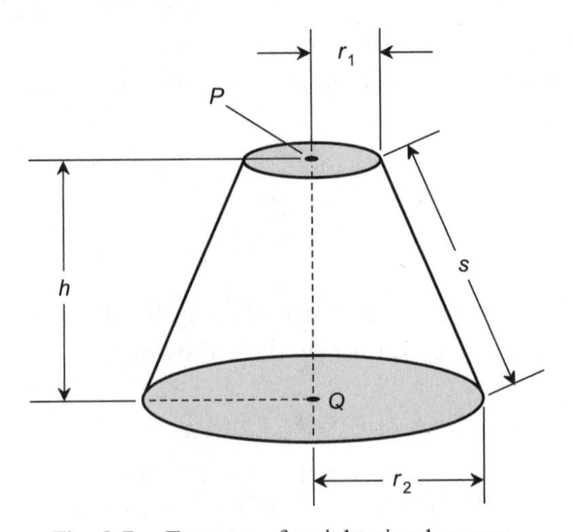

Fig. 8-7. Frustum of a right circular cone.

VOLUME OF FRUSTUM OF RIGHT CIRCULAR CONE

Imagine a frustum of a right circular cone as defined above and in Fig. 8-7. The volume, V, of the interior of the object is given by this formula:

$$V = \pi h(r_1^2 + r_1 r_2 + r_2^2)/3$$

THE SLANT CIRCULAR CONE

A *slant circular cone* has a base that is a circle, and an apex point such that a normal line from the apex point to the plane of the base does not pass through the center of the base (Fig. 8-8).

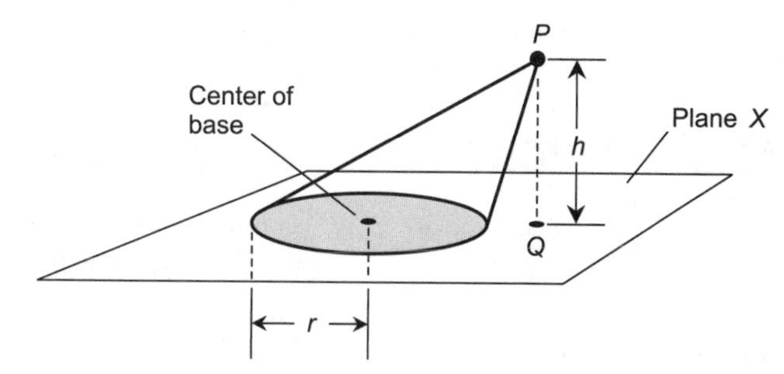

Fig. 8-8. A slant circular cone.

VOLUME OF SLANT CIRCULAR CONE

Imagine a cone whose base is a circle. Let P be the apex of the cone, and let Q be a point in the plane X containing the base such that line segment PQ is perpendicular to X (Fig. 8-8). Let h be the height of the cone (the length of line segment PQ). Let r be the radius of the base. Then the volume, V, of the corresponding cone is given by:

$$V = \pi r^2 h/3$$

This is the same as the formula for the volume of a right circular cone.

THE RIGHT CIRCULAR CYLINDER

A *right circular cylinder* has a circular base and a circular top. The base and the top lie in parallel planes. The center of the base and the center of the top lie along a line that is normal to both the plane containing the base and the plane containing the top (Fig. 8-9).

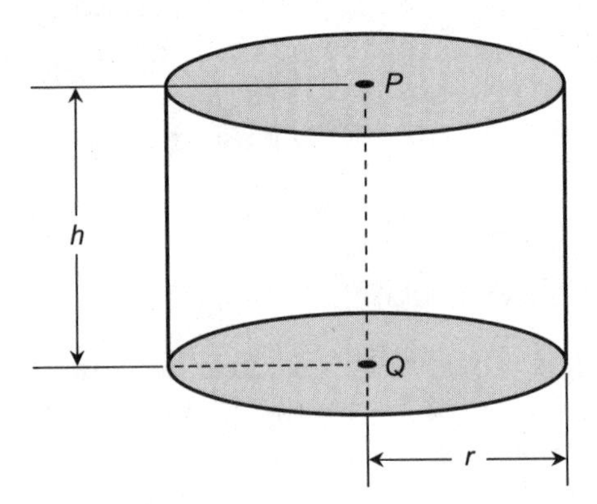

Fig. 8-9. A right circular cylinder.

SURFACE AREA OF RIGHT CIRCULAR CYLINDER

Imagine a right circular cylinder where P is the center of the top and Q is the center of the base (Fig. 8-9). Let r be the radius of the cylinder, and let h be the height (the length of line segment PQ). Then the surface area S_1 of the cylinder, including the base and the top, is given by:

$$S_1 = 2\pi rh + 2\pi r^2 = 2\pi r\ (h + r)$$

The lateral surface area S_2 (not including the base or the top) is given by:

$$S_2 = 2\pi rh$$

VOLUME OF RIGHT CIRCULAR CYLINDER

Imagine a right circular cylinder as defined above and shown in Fig. 8-9. The volume, V, of the cylinder is given by:

$$V = \pi r^2 h$$

THE SLANT CIRCULAR CYLINDER

A *slant circular cylinder* has a circular base and a circular top. The base and the top lie in parallel planes. The center of the base and the center of the top lie along a line that is not perpendicular to the planes that contain them (Fig. 8-10).

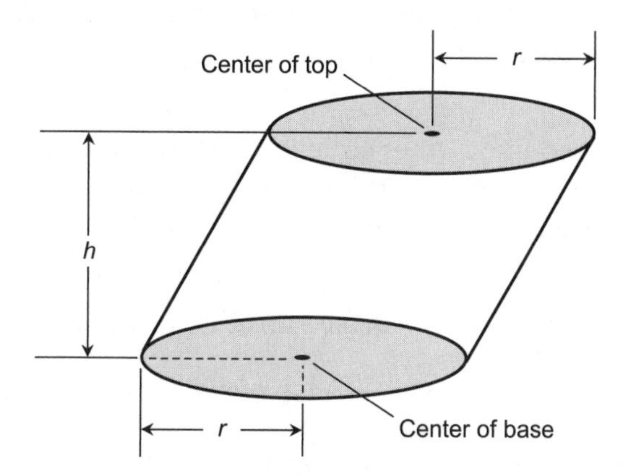

Fig. 8-10. A slant circular cylinder.

VOLUME OF SLANT CIRCULAR CYLINDER

Imagine a slant circular cylinder as defined above and in Fig. 8-10. The volume, V, of the corresponding solid is given by:

$$V = \pi r^2 h$$

PROBLEM 8-2

A cylindrical water tower is exactly 30 meters high and exactly 10 meters in radius. How many liters of water can it hold, assuming the entire interior can be filled with water? (One liter is equal to a cubic decimeter, or the volume of a cube measuring 0.1 meters on an edge.) Round the answer off to the nearest liter.

SOLUTION 8-2

Use the formula for the volume of a right circular cylinder to find the volume in cubic meters:

$$V = \pi r^2 h$$

Plugging in the numbers, let $r = 10$, $h = 30$, and $\pi = 3.14159$:

$$\begin{aligned} V &= 3.14159 \times 10^2 \times 30 \\ &= 3.14159 \times 100 \times 30 \\ &= 9424.77 \end{aligned}$$

One liter is the volume of a cube measuring 10 centimeters, or 0.1 meter, on an edge (believe it or not!). Thus, there are 1000 liters in a cubic meter. This means that the amount of water the tower can hold, in liters, is equal to 9424.77×1000, or 9,424,770.

PROBLEM 8-3

A circus tent is shaped like a right circular cone. Its diameter is 50 meters and the height at the center is 20 meters. How much canvas is in the tent? Express the answer to the nearest square meter.

SOLUTION 8-3

Use the formula for the lateral surface area, S, of the right circular cone:

$$S = \pi r(r^2 + h^2)^{1/2}$$

We know that the diameter is 50 meters, so the radius is 25 meters. Therefore, $r = 25$. We also know that $h = 20$. Let $\pi = 3.14159$. Then:

$$\begin{aligned} S &= 3.14159 \times 25 \times (25^2 + 20^2)^{1/2} \\ &= 3.14159 \times 25 \times (625 + 400)^{1/2} \\ &= 3.14159 \times 25 \times 1025^{1/2} \\ &= 3.14159 \times 25 \times 32.0156 \\ &= 2514.4972201 \end{aligned}$$

There are 2514 square meters of canvas, rounded off to the nearest square meter.

Other Solids

There exists an incredible variety of geometric solids that have curved surfaces throughout. Here, we'll look at three of the most common: the *sphere*, the *ellipsoid*, and the *torus*.

THE SPHERE

Consider a specific point P in 3D space. The surface of a sphere S consists of the set of all points at a specific distance or radius r from point P. The interior of sphere S, including the surface, consists of the set of all points whose distance from point P is less than or equal to r. The interior of sphere S, not including the surface, consists of the set of all points whose distance from P is less than r.

SURFACE AREA OF SPHERE

Imagine a sphere S having radius r as shown in Fig. 8-11. The surface area, A, of the sphere is given by:

$$A = 4\pi r^2$$

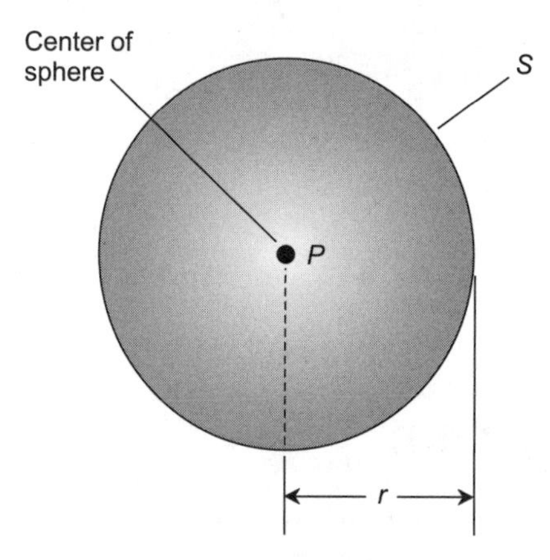

Fig. 8-11. A sphere.

VOLUME OF SPHERE

Imagine a sphere S as defined above and in Fig. 8-11. The volume, V, of the solid enclosed by the sphere is given by:

$$V = 4\pi r^3/3$$

This volume applies to the interior of sphere S, either including the surface or not including it, because the surface has zero volume.

THE ELLIPSOID

Let E be a set of points that forms a closed surface. Then E is an ellipsoid if and only if, for any plane X that intersects E, the intersection between E and X is either a single point, a circle, or an ellipse.

Figure 8-12 shows an ellipsoid E with center point P and radii r_1, r_2, and r_3 as specified in a 3D rectangular coordinate system with P at the origin. If r_1, r_2, and r_3 are all equal, then E is a sphere, which is a special case of the ellipsoid.

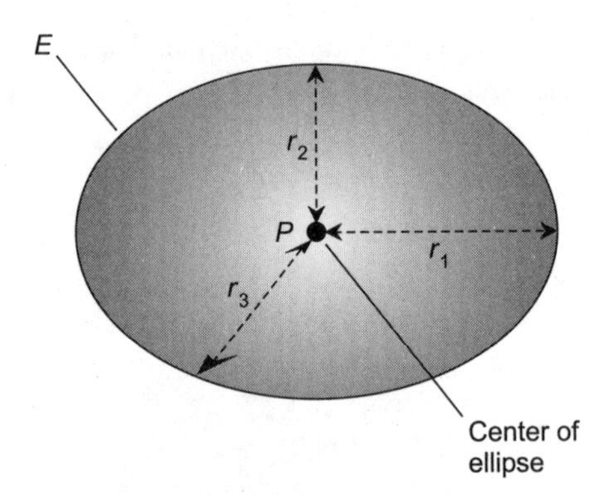

Fig. 8-12. An ellipsoid.

VOLUME OF ELLIPSOID

Imagine an ellipsoid whose semi-axes are r_1, r_2, and r_3 (Fig. 8-12). The volume, V, of the enclosed solid is given by:

$$V = 4\pi r_1 r_2 r_3 / 3$$

THE TORUS

Imagine a ray PQ, and a small circle C centered on point Q whose radius is less than half of the distance between points P and Q. Suppose ray PQ, along with the small circle C centered at point Q, is rotated around its end point, P, so that point Q describes a circle that lies in a plane perpendicular to the small circle C. The resulting set of points in 3D space, "traced out" by circle C, is a torus.

Figure 8-13 shows a torus T thus constructed, with center point P. The inside radius is r_1 and the outside radius is r_2. The torus is sometimes informally called a "donut."

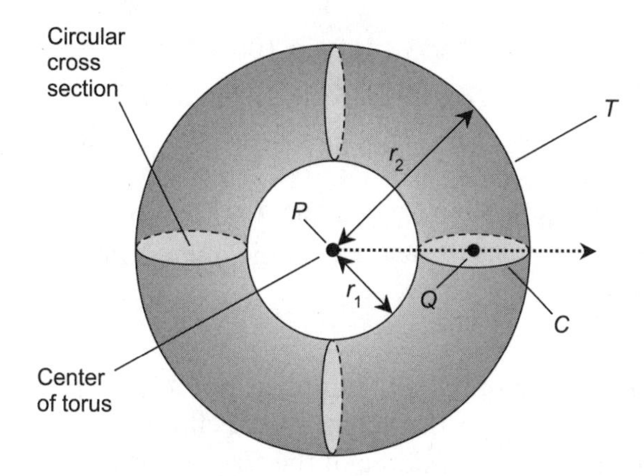

Fig. 8-13. A torus, also called a "donut."

SURFACE AREA OF TORUS

Imagine a torus with an inner radius of r_1 and an outer radius of r_2 as shown in Fig. 8-13. The surface area, A, of the torus is given by:

$$A = \pi^2(r_2 + r_1)(r_2 - r_1)$$

VOLUME OF TORUS

Let T be a torus as defined above and in Fig. 8-13. The volume, V, of the enclosed solid is given by:

$$V = \pi^2(r_2 + r_1)(r_2 - r_1)^2/4$$

PROBLEM 8-4

Suppose a football field is to be covered by an inflatable dome that takes the shape of a half-sphere. If the radius of the dome is 100 meters, what is the volume of air enclosed by the dome in cubic meters? Find the result to the nearest 1000 cubic meters.

SOLUTION 8-4

First, find the volume V of a sphere whose radius is 100 meters, and then divide the result by 2. Let $\pi = 3.14159$. Using the formula with $r = 100$ gives this result:

$$\begin{aligned} V &= (4 \times 3.14159 \times 100^3)/3 \\ &= (4 \times 3.14159 \times 1{,}000{,}000)/3 \\ &= 4{,}188{,}786.667\ldots \end{aligned}$$

Thus $V/2 = 4{,}188{,}786.667/2 = 2{,}094{,}393.333$. Rounding off to the nearest 1000 cubic meters, we get 2,094,000 cubic meters as the volume of air enclosed by the dome.

PROBLEM 8-5

Suppose the dome in the previous example is not a half-sphere, but instead is a half-ellipsoid. Imagine that the height of the ellipsoid is 70 meters above its center point, which lies in the middle of the 50-yard line at field level. Suppose that the distance from the center of the 50-yard line to either end of the dome, as measured parallel to the sidelines, is 120 meters, and the distance from the center of the 50-yard line, as measured along the line containing the 50-yard line itself, is 90 meters. What is the volume of air, to the nearest 1000 cubic meters, enclosed by this dome?

SOLUTION 8-5

First, consider the radii r_1, r_2, and r_3 in meters, with respect to the center point, as follows:

$$\begin{aligned} r_1 &= 120 \\ r_2 &= 90 \\ r_3 &= 70 \end{aligned}$$

Then use the formula for the volume V of an ellipsoid with these radii:

$$\begin{aligned} V &= (4 \times 3.14159 \times 120 \times 90 \times 70)/3 \\ &= (4 \times 3.14159 \times 756{,}000)/3 \\ &= 3{,}166{,}722.72 \end{aligned}$$

Thus $V/2 = 3{,}166{,}722.72/2 = 1{,}583{,}361.36$. Rounding off to the nearest 1000 cubic meters, we get 1,583,000 cubic meters as the volume of air enclosed by the half-ellipsoidal dome.

Quiz

Refer to the text in this chapter if necessary. A good score is eight correct. Answers are in the back of the book.

1. What is the approximate volume of a circular cone whose base has an area of 30 square units and a height of 10 units?
   (a) 100 cubic units
   (b) 150 cubic units
   (c) 300 cubic units
   (d) There is not enough information given here to calculate it
2. Consider the earth to be a perfect sphere 12,800 kilometers in diameter. What is the approximate surface area of the earth based on this figure?
   (a) 515 million square kilometers
   (b) 2059 million square kilometers
   (c) 1.1 trillion square kilometers
   (d) 8.8 trillion square kilometers
3. What is the set of points representing the intersection of a torus with a plane containing its center?
   (a) A pair of non-concentric circles
   (b) A pair of concentric circles
   (c) A pair of non-concentric ellipses
   (d) More information is needed to answer this question
4. A rectangular prism has
   (a) six edges, all of which are the same length
   (b) eight edges of various lengths
   (c) six faces, all of which are the same shape
   (d) none of the above
5. If all other factors are held constant, the surface area of a torus depends on all of the following except
   (a) its inner radius
   (b) its outer radius
   (c) the difference between the squares of its inner and outer radii
   (d) its orientation in space
6. If all other factors are held constant, the volume of a parallelepiped depends on
   (a) the height
   (b) the width
   (c) the ratio of the height to the width
   (d) more than one of the above
7. In a circular cylinder, the height is equal to
   (a) the radius of either the base or the top multiplied by 2π

(b) the distance between the planes containing the base and the top
(c) the distance between the center of the base and the center of the top
(d) the distance along any straight line in the object's periphery

8. The volume of a rectangular prism is equal to
(a) the sum of the lengths of its edges
(b) the product of the lengths of its edges
(c) the sum of the surface areas of its faces
(d) the product of the surface areas of its faces

9. The faces of a tetrahedron
(a) are all triangles
(b) are all quadrilaterals
(c) are all triangles or quadrilaterals
(d) all lie in the same plane

10. Imagine a cube with edges measuring 10 meters each. Suppose a pyramid is carved from this cube, such that the base of the pyramid corresponds to one of the faces of the cube, and the apex of the pyramid is at the center of the face of the cube opposite the base. What is the approximate volume of the pyramid?
(a) 1000 cubic meters
(b) 707 cubic meters
(c) 333 cubic meters
(d) There is not enough information given here to calculate it

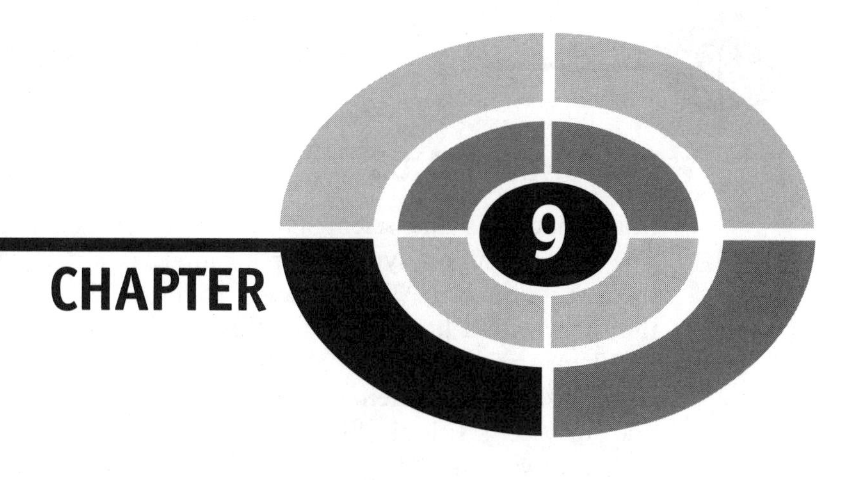

Vectors and Cartesian Three-Space

Cartesian three-space, also called *rectangular three-space* or *xyz-space*, is defined by three number lines that intersect at a common origin point. At the origin, each of the three number lines is perpendicular to the other two. This makes it possible to pictorially relate one variable to another. Most three-dimensional (3D) graphs look like lines, curves, or surfaces. Renditions are enhanced by computer graphics programs.

The approach here is similar to that of Chapter 6. You will need to know middle-school algebra to understand the material in this chapter.

A Taste of Trigonometry

Before we proceed further, let's get familiar with some basic trigonometry. In particular, let's look at angle notation and the *sine*, *cosine*, and *tangent* functions.

IT'S GREEK TO US

Mathematicians and scientists often use Greek letters to represent angles. The most common symbol for an angle is an italicized, lowercase Greek theta (pronounced "THAY-tuh"). It looks like a numeral zero leaning to the right, with a horizontal line through it (θ).

When writing about two different angles, a second Greek letter is used along with θ. Most often, it is the italicized, lowercase letter phi (pronounced "fie" or "fee"). It looks like a lowercase English letter o leaning to the right, with a forward slash through it (ϕ). Numeric or variable subscripts are sometimes used along with the Greek symbols, so don't be surprised if you see angles denoted θ_1, θ_2, θ_3 or θ_x, θ_y, θ_z.

THE UNIT CIRCLE

Consider a circle in the Cartesian *xy*-plane with the following equation:

$$x^2 + y^2 = 1$$

This equation represents a *unit circle* because it is centered at the origin and has a radius of one unit. Let θ be an angle whose apex is at the origin, and that is measured counterclockwise from the x axis, as shown in Fig. 9-1. Suppose this angle corresponds to a ray that intersects the unit circle at

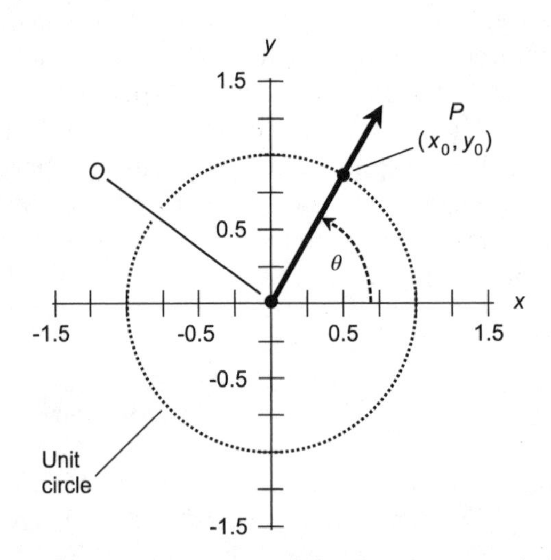

Fig. 9-1. The unit circle is the basis for the trigonometric functions.

some point $P = (x_0, y_0)$. Then we can define three mathematical functions, called *circular functions*, of the angle θ in a simple way.

THE SINE FUNCTION

In Fig. 9-1, let ray OP be defined as the ray from the origin (point O) passing outward through point P on the unit circle. Imagine this ray at first pointing right along the x axis, and then turning around and around in a counter-clockwise direction. As the ray turns, the point P, represented by coordinates (x_0, y_0), revolves around the unit circle.

Imagine what happens to the value of y_0 during one complete revolution of the ray: it starts out at $y_0 = 0$, then increases until it reaches $y_0 = 1$ after P has gone 90° or $\pi/2$ rad around the circle ($\theta = 90° = \pi/2$ rad). After that, y_0 begins to decrease, getting back to $y_0 = 0$ when P has gone 180° or π rad around the circle ($\theta = 180° = \pi$ rad). As P continues on its counterclockwise trek, y_0 keeps decreasing until, at $\theta = 270° = 3\pi/2$ rad, the value of y_0 reaches its minimum of –1. After that, the value of y_0 rises again until, when P has gone completely around the circle, it returns to $y_0 = 0$ for $\theta = 360° = 2\pi$ rad.

The value of y_0 is defined as the sine of the angle θ. The sine function is abbreviated sin, so we can state this simple equation:

$$\sin \theta = y_0$$

THE COSINE FUNCTION

Look again at Fig. 9-1. Imagine, once again, a ray OP from the origin outward through point P on the circle, pointing right along the x axis, and then rotating in a counterclockwise direction.

What happens to the value of x_0 during one complete revolution of the ray? It starts out at $x_0 = 1$, then decreases until it reaches $x_0 = 0$ when $\theta = 90° = \pi/2$ rad. After that, x_0 continues to decrease, getting down to $x_0 = -1$ when $\theta = 180° = \pi$ rad. As P continues counterclockwise around the circle, x_0 begins to increase again; at $\theta = 270° = 3\pi/2$ rad, the value gets back up to $x_0 = 0$. After that, x_0 increases further until, when P has gone completely around the circle, it returns to $x_0 = 1$ for $\theta = 360° = 2\pi$ rad.

The value of x_0 is defined as the cosine of the angle θ. The cosine function is abbreviated cos. So we can write this:

$$\cos \theta = x_0$$

THE TANGENT FUNCTION

Once again, refer to Fig. 9-1. The tangent (abbreviated tan) of an angle θ is defined using the same ray OP and the same point $P = (x_0, y_0)$ as is done with the sine and cosine functions. The definition is:

$$\tan \theta = y_0 / x_0$$

Because we already know that $\sin \theta = y_0$ and $\cos \theta = x_0$, we can express the tangent function in terms of the sine and the cosine:

$$\tan \theta = \sin \theta / \cos \theta$$

This function is interesting because, unlike the sine and cosine functions, it "blows up" at certain values of θ. Whenever $x_0 = 0$, the denominator of either quotient above becomes zero. Division by zero is not defined, and that means the tangent function is not defined for any angle θ such that $\cos \theta = 0$. Such angles are all the odd multiples of 90° ($\pi/2$ rad).

PROBLEM 9-1

What is tan 45°? Do not perform any calculations. You should be able to infer this without having to write down a single numeral, and without using a calculator.

SOLUTION 9-1

Draw a diagram of a unit circle, such as the one in Fig. 9-1, and place ray OP such that it subtends an angle of 45° with respect to the x axis. That angle is the angle of which we want to find the tangent. Note that the ray OP also subtends an angle of 45° with respect to the y axis, because the x and y axes are perpendicular (they are oriented at 90° with respect to each other), and 45° is exactly half of 90°. Every point on the ray OP is equally distant from the x and y axes; this includes the point (x_0, y_0). It follows that x_0 and y_0 must be the same, and neither of them is zero. From this, we can conclude that $y_0/x_0 = 1$. According to the definition of the tangent function, therefore, $\tan 45° = 1$.

Vectors in the Cartesian Plane

A *vector* is a mathematical expression for a quantity with two independent properties: *magnitude* and *direction*. The direction, also called *orientation*, is defined in the sense of a ray, so it "points" somewhere. Vectors are used to represent physical variables such as distance, velocity, and acceleration.

Conventionally, vectors are denoted by boldface letters of the alphabet. In the xy-plane, vectors **a** and **b** can be illustrated as rays from the origin (0,0) to points (x_a,y_a) and (x_b,y_b) as shown in Fig. 9-2.

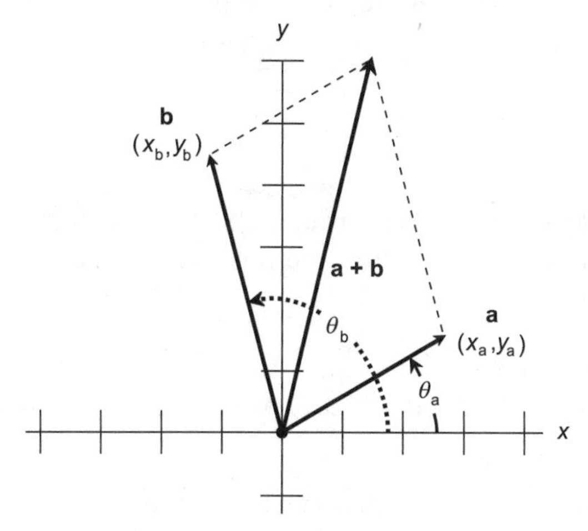

Fig. 9-2. Two vectors in the Cartesian plane. They are added using the "parallelogram method."

EQUIVALENT VECTORS

Occasionally, a vector is expressed in a form that begins at a point other than the origin (0,0). In order for the following formulas to hold, such a vector must be reduced to so-called *standard form*, where it begins at the origin. This can be accomplished by subtracting the coordinates (x_0,y_0) of the starting point from the coordinates of the end point (x_1,y_1). For example, if a vector **a*** starts at (3,–2) and ends at (1,–3), it reduces to an *equivalent vector* **a** in standard form:

$$\begin{aligned}\mathbf{a} &= \{(1-3), [-3-(-2)]\}\\ &= (-2, -1)\end{aligned}$$

Any vector **a*** that is parallel to **a**, and that has the same length and the same direction (or orientation) as **a**, is equal to vector **a**. A vector is defined solely on the basis of its magnitude and its direction (or orientation). Neither of these two properties depends on the location of the end point.

MAGNITUDE

The magnitude (also called the *length*, *intensity*, or *absolute value*) of vector **a**, written |**a**| or *a*, can be found in the Cartesian plane by using a distance formula resembling the Pythagorean theorem:

$$|\mathbf{a}| = (x_a^2 + y_a^2)^{1/2}$$

DIRECTION

The direction of vector **a**, written dir **a**, is the angle θ_a that vector **a** subtends as expressed counterclockwise from the positive x axis:

$$\text{dir } \mathbf{a} = \theta_a$$

The tangent of the angle θ_a is equal to y_a/x_a. Therefore, θ_a is equal to the *inverse tangent*, also called the *arctangent* (abbreviated arctan or $\tan^{-1}$) of y_a/x_a. Therefore:

$$\text{dir } \mathbf{a} = \theta_a = \arctan\,(y_a/x_a) = \tan^{-1}(y_a/x_a)$$

By convention, the angle θ_a is reduced to a value that is at least zero, but less than one full counterclockwise revolution. That is, $0° \leq \theta_a < 360°$ (if the angle is expressed in degrees), or $0 \text{ rad} \leq \theta_a < 2\pi \text{ rad}$ (if the angle is expressed in radians).

SUM

The sum of two vectors **a** and **b**, where $\mathbf{a} = (x_a, y_a)$ and $\mathbf{b} = (x_b, y_b)$, is given by the following formula:

$$\mathbf{a} + \mathbf{b} = [(x_a + x_b), (y_a + y_b)]$$

This sum can be found geometrically by constructing a parallelogram with **a** and **b** as adjacent sides. Then **a** + **b** is the diagonal of this parallelogram (Fig. 9-2).

MULTIPLICATION BY SCALAR

To multiply a vector by a *scalar* (an ordinary real number), the x and y components of the vector are both multiplied by that scalar. If we have a vector $\mathbf{a} = (x_a, y_a)$ and a scalar k, then

$$k\mathbf{a} = \mathbf{a}k = (kx_a, ky_a)$$

Multiplication by a scalar changes the length of a vector. If the scalar is positive, the direction of the product vector is the same as that of the original vector. If the scalar is negative, the direction of the product vector is opposite that of the original vector. If the scalar is zero, the product vector vanishes.

DOT PRODUCT

Let $\mathbf{a} = (x_a, y_a)$ and $\mathbf{b} = (x_b, y_b)$. The *dot product*, also known as the *scalar product* and written $\mathbf{a} \bullet \mathbf{b}$, of two vectors **a** and **b** is a real number (that is, a scalar) given by the formula:

$$\mathbf{a} \bullet \mathbf{b} = x_a x_b + y_a y_b$$

PROBLEM 9-2

What is the sum of $\mathbf{a} = (3, -5)$ and $\mathbf{b} = (2,6)$?

SOLUTION 9-2

Add the x and y components together independently:

$$\begin{aligned}\mathbf{a} + \mathbf{b} &= [(3 + 2), (-5 + 6)] \\ &= (5,1)\end{aligned}$$

PROBLEM 9-3

What is the dot product of $\mathbf{a} = (3, -5)$ and $\mathbf{b} = (2,6)$?

SOLUTION 9-3

Use the formula given above for the dot product:

$$\begin{aligned}\mathbf{a} \bullet \mathbf{b} &= (3 \times 2) + (-5 \times 6) \\ &= 6 + (-30) \\ &= -24\end{aligned}$$

PROBLEM 9-4

What happens if the order of the dot product is reversed? Does the value change?

SOLUTION 9-4

No. The value of the dot product of two vectors does not depend on the order in which the vectors are "dot-multiplied." This can be proven in the general case using the formula above. Let $\mathbf{a} = (x_a, y_a)$ and $\mathbf{b} = (x_b, y_b)$. First consider the dot product of **a** and **b** (pronounced "**a** dot **b**"):

$$\mathbf{a} \bullet \mathbf{b} = x_a x_b + y_a y_b$$

Now consider the dot product **b** • **a**:

$$\mathbf{b} \bullet \mathbf{a} = x_b x_a + y_b y_a$$

Because ordinary multiplication is commutative – that is, the order in which the factors are multiplied doesn't matter – we can convert the above formula to this:

$$\mathbf{b} \bullet \mathbf{a} = x_a x_b + y_a y_b$$

But $x_a x_b + y_a y_b$ is the expansion of **a** • **b**. Therefore, for any two vectors **a** and **b**, it is always true that **a** • **b** = **b** • **a**.

Three Number Lines

Figure 9-3 illustrates the simplest possible set of *rectangular 3D coordinates*. All three number lines have equal increments. (This is a perspective illustration, so the increments on the *z* axis appear distorted. A true 3D rendition would have the positive *z* axis perpendicular to the page.) The three number lines intersect at their zero points.

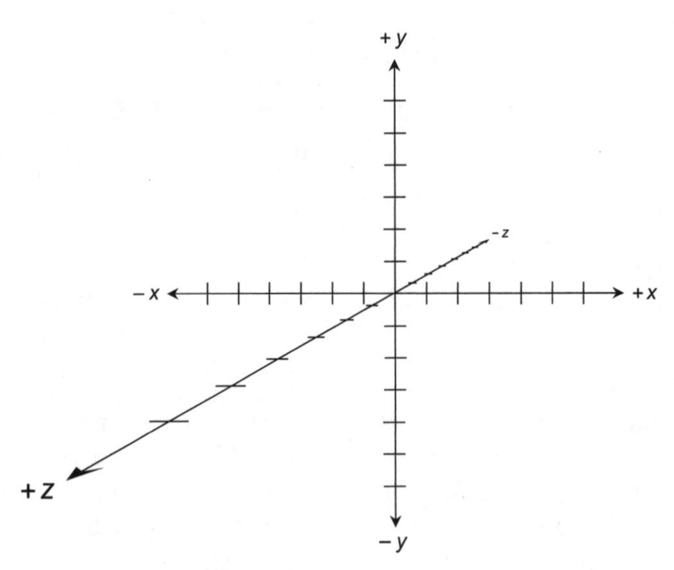

Fig. 9-3. Cartesian three-space, also called *xyz*-space.

The horizontal (right-and-left) axis is called the x axis; the vertical (up-and-down) axis is called the y axis, and the page-perpendicular (in-and-out) axis is called the z axis. In most renditions of rectangular 3D coordinates, the positive x axis runs from the origin toward the viewer's right, and the negative x axis runs toward the left. The positive y axis runs upward, and the negative y axis runs downward. The positive z axis comes "out of the page," and the negative z axis extends "back behind the page."

ORDERED TRIPLES AS POINTS

Figure 9-4 shows two specific points, called P and Q, plotted in Cartesian three-space. The coordinates of point P are (–5,–4,3), and the coordinates of point Q are (3,5,–2). Points are denoted as *ordered triples* in the form (x,y,z), where the first number represents the value on the x axis, the second number represents the value on the y axis, and the third number represents the value on the z axis. The word "ordered" means that the order, or sequence, in which the numbers are listed is important. The ordered triple (1,2,3) is not the same as any of the ordered triples (1,3,2), (2,1,3), (2,3,1), (3,1,2), or (3,2,1), even though all of the triples contain the same three numbers.

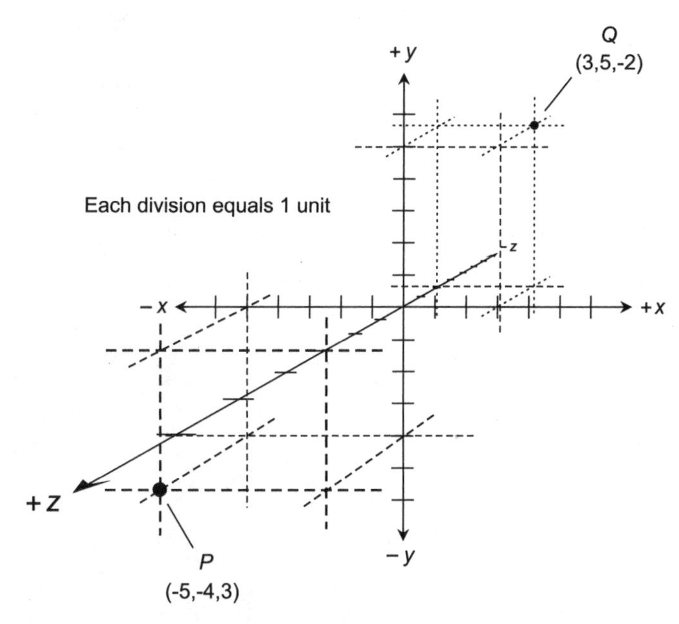

Fig. 9-4. Two points in Cartesian three-space.

In an ordered triple, there are no spaces after the commas, as there are in the notation of a set or sequence. The rule is the same as that for ordered pairs.

VARIABLES AND ORIGIN

In Cartesian three-space, there are usually two independent-variable coordinate axes and one dependent-variable axis. The x and y axes represent independent variables; the z axis represents a dependent variable whose value is affected by both the x and the y values.

In some scenarios, two of the variables are dependent and only one is independent. Most often, the independent variable in such cases is x. Rarely, you'll come across a situation in which none of the values depends on either of the other two, or when a correlation, but not a true mathematical relation, exists among the values of two or all three of the variables. Plots of this sort usually look like "swarms of points," representing the results of observations, or values predicted by some scientific theory.

DISTANCE BETWEEN POINTS

Suppose there are two different points $P = (x_0,y_0,z_0)$ and $Q = (x_1,y_1,z_1)$ in Cartesian three-space. The distance d between these two points can be found using this formula:

$$d = [(x_1 - x_0)^2 + (y_1 - y_0)^2 + (z_1 - z_0)^2]^{1/2}$$

PROBLEM 9-5

What is the distance between the points $P = (-5,-4,3)$ and $Q = (3,5,-2)$ illustrated in Fig. 9-4? Express the answer rounded off to three decimal places.

SOLUTION 9-5

We can plug the coordinate values into the distance equation, where:

$$x_0 = -5$$
$$x_1 = 3$$
$$y_0 = -4$$
$$y_1 = 5$$
$$z_0 = 3$$
$$z_1 = -2$$

Therefore:

$$\begin{aligned} d &= \{[3-(-5)]^2 + [5-(-4)]^2 + (-2-3)^2\}^{1/2} \\ &= [8^2 + 9^2 + (-5)^2]^{1/2} \\ &= (64 + 81 + 25)^{1/2} \\ &= 170^{1/2} \\ &= 13.038 \end{aligned}$$

Vectors in Cartesian Three-Space

A *vector* in Cartesian three-space is the same as a vector in the Cartesian plane, except that there is more "freedom" in terms of direction. This makes the expression of direction in 3D more complicated than is the case in 2D. It also makes vector arithmetic a lot more interesting!

EQUIVALENT VECTORS

In Cartesian three-space, vectors **a** and **b** can be denoted as arrow-tipped line segments from the origin (0,0,0) to points (x_a,y_a,z_a) and (x_b,y_b,z_b), as shown in Fig. 9-5. This, like all three-space drawings in this chapter, is a perspective illustration. Both vectors in this example point in directions on the reader's side of the plane containing the page. In a true 3D model, both of them would "stick up out of the paper at an angle."

In Fig. 9-5, both vectors **a** and **b** have their end points at the origin. This is the standard form of a vector in any coordinate system. In order for the following formulas to hold, vectors must be expressed in standard form. If a given vector is not in standard form, it can be converted by subtracting the coordinates (x_0,y_0,z_0) of the starting point from the coordinates of the end point (x_1,y_1,z_1). For example, if a vector **a*** starts at (4,7,0) and ends at (1,–3,5), it reduces to an equivalent vector **a** in standard form:

$$\begin{aligned} \mathbf{a} &= [(1-4), (-3-7), (5-0)] \\ &= (-3, -10, 5) \end{aligned}$$

Any vector **a***, which is parallel to **a** and has the same length as **a**, is equal to vector **a**, because **a*** has the same magnitude and the same direction as **a**. Similarly, any vector **b***, which is parallel to **b** and has the same length as **b**, is

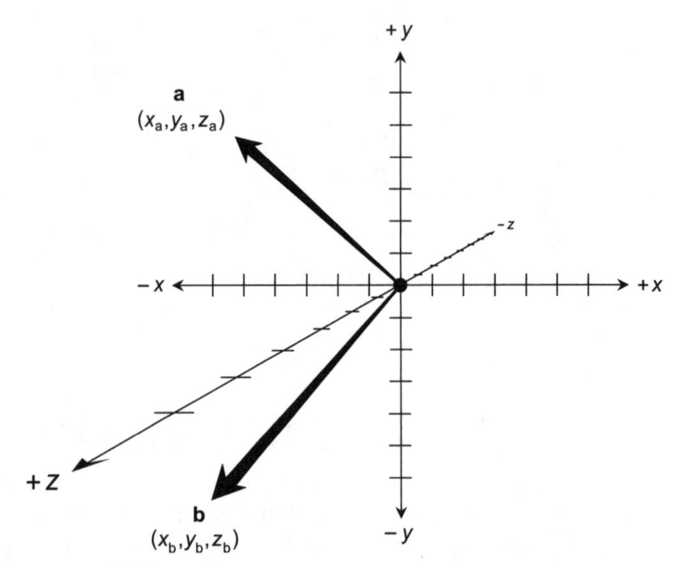

Fig. 9-5. Vectors in *xyz*-space. This is a perspective drawing; both vectors point in directions on your side of the plane containing the page.

defined as being equal to **b**. As in the 2D case, a vector is defined solely on the basis of its magnitude and its direction. Neither of these two properties depends on the location of the end point.

DEFINING THE MAGNITUDE

When the end point of a vector **a** is at the origin, the magnitude of **a**, written $|\mathbf{a}|$ or a, can be found by a three-dimensional extension of the Pythagorean theorem for right triangles. The formula looks like this:

$$|\mathbf{a}| = (x_a^2 + y_a^2 + z_a^2)^{1/2}$$

The magnitude of any vector **a** in standard form is simply the distance of the end point from the origin. Note that the above formula is the distance formula for two points, (0,0,0) and (x_a,y_a,z_a).

DIRECTION ANGLES AND COSINES

The direction of a vector **a** in standard form can be defined by specifying the angles θ_x, θ_y, and θ_z that the vector **a** subtends relative to the positive x, y, and z axes respectively (Fig. 9-6). These angles, expressed in radians as an ordered triple $(\theta_x,\theta_y,\theta_z)$, are the *direction angles* of **a**.

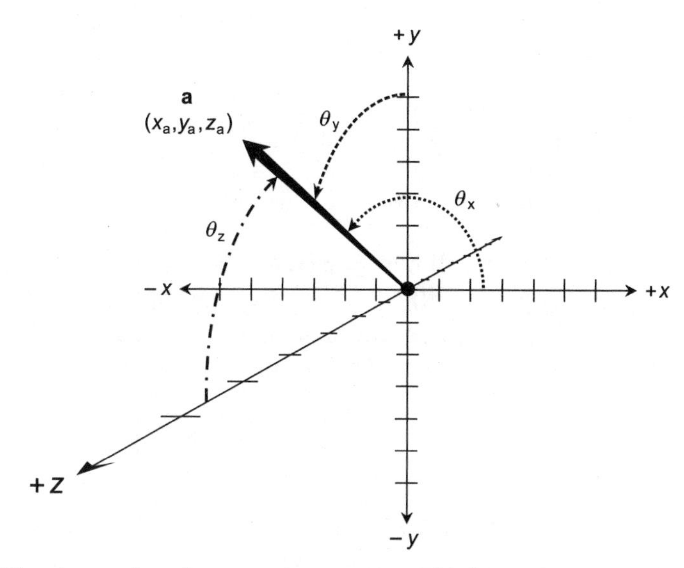

Fig. 9-6. Direction angles of a vector in *xyz*-space. This is another perspective drawing; the vector points in a direction on your side of the plane containing the page.

Sometimes the cosines of these angles are used to define the direction of a vector **a** in 3D space. These are the *direction cosines* of **a**:

$$\text{dir } \mathbf{a} = (\alpha,\beta,\gamma)$$

$$\alpha = \cos \theta_x$$

$$\beta = \cos \theta_y$$

$$\gamma = \cos \theta_z$$

For any vector **a** in Cartesian three-space, the sum of the squares of the direction cosines is always equal to 1. That is

$$\alpha^2 + \beta^2 + \gamma^2 = 1$$

Another way of expressing this is:

$$\cos^2\theta_x + \cos^2 \theta_y + \cos^2 \theta_z = 1$$

where the expression $\cos^2 \theta$ means $(\cos \theta)^2$.

SUM

The sum of vectors $\mathbf{a} = (x_a,y_a,z_a)$ and $\mathbf{b} = (x_b,y_b,z_b)$ in standard form is given by the following formula:

$$\mathbf{a} + \mathbf{b} = [(x_a + x_b), (y_a + y_b), (z_a + z_b)]$$

This sum can, as in the two-dimensional case, be found geometrically by constructing a parallelogram with **a** and **b** as adjacent sides. The sum **a** + **b** is the diagonal of the parallelogram. This is shown in Fig. 9-7. (The parallelogram appears distorted because of the perspective of the drawing.)

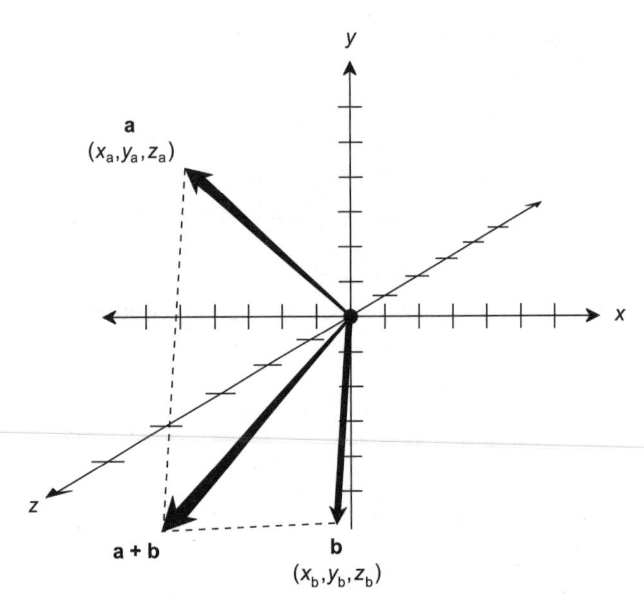

Fig. 9-7. Vectors in *xyz*-space are added using the "parallelogram method." This is a perspective drawing, so the parallelogram appears distorted.

MULTIPLICATION BY SCALAR

In three-dimensional Cartesian coordinates, let vector **a** be defined by the coordinates (x_a,y_a,z_a) when reduced to standard form. Suppose **a** is multiplied by a positive real scalar k. Then the following equation holds:

$$k\mathbf{a} = k(x_a,y_a,z_a) = (kx_a,ky_a,kz_a)$$

If **a** is multiplied by a negative real scalar $-k$, then:

$$-k\mathbf{a} = -k(x_a,y_a,z_a) = (-kx_a, -ky_a, -kz_a)$$

Suppose the direction angles of **a** are represented by the ordered triple $(\theta_{xa},\theta_{ya},\theta_{za})$. Then the direction angles of k**a** are the same; they are also $(\theta_{xa},\theta_{ya},\theta_{za})$. The direction angles of $-k$**a** are all changed by 180° (π rad). The direction angles of $-k$**a** are obtained by adding or subtracting 180° (π rad) to each of the direction angles for k**a**, so that the results are all at least 0° (0 rad) but less than 360° (2π rad).

DOT PRODUCT

The *dot product*, also known as the *scalar product* and written **a** • **b**, of vectors $\mathbf{a} = (x_a,y_a,z_a)$ and $\mathbf{b} = (x_b,y_b,z_b)$ in standard form is a real number given by the formula:

$$\mathbf{a} \bullet \mathbf{b} = x_a x_b + y_a y_b + z_a z_b$$

The dot product can also be found from the magnitudes $|\mathbf{a}|$ and $|\mathbf{b}|$, and the angle θ between vectors **a** and **b** as measured counterclockwise in the plane containing them both:

$$\mathbf{a} \bullet \mathbf{b} = |\mathbf{a}||\mathbf{b}| \cos \theta$$

CROSS PRODUCT

The *cross product*, also known as the *vector product* and written **a** × **b**, of vectors $\mathbf{a} = (x_a,y_a,z_a)$ and $\mathbf{b} = (x_b,y_b,z_b)$ in standard form is a vector perpendicular to the plane containing **a** and **b**. Let θ be the angle between vectors **a** and **b** as measured counterclockwise in the plane containing them both, as shown in Fig. 9-8. The magnitude of **a** × **b** is given by the formula:

$$|\mathbf{a} \times \mathbf{b}| = |\mathbf{a}||\mathbf{b}| \sin \theta$$

In the example shown, **a** × **b** points upward at a right angle to the plane containing both vectors **a** and **b**. If $0° < \theta < 180°$ ($0 < \theta < \pi$), you can use the *right-hand rule* to ascertain the direction of **a** × **b**. Curl your fingers in the direction that θ, the angle between **a** and **b**, is defined. Extend your thumb. Then **a** × **b** points in the direction of your thumb.

When $180° < \theta < 360°$ (π rad $< \theta < 2\pi$ rad), the cross-product vector reverses direction because its magnitude becomes negative. This is demonstrated by the fact that, in the above formula, $\sin \theta$ is positive when $0° < \theta < 180°$ (0 rad $< \theta < \pi$ rad), but negative when $180° < \theta < 360°$ (π rad $< \theta < 2\pi$ rad).

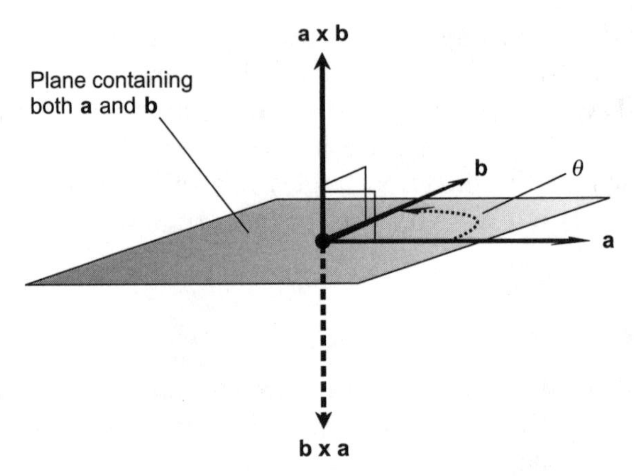

Fig. 9-8. The vector $\mathbf{b} \times \mathbf{a}$ has the same magnitude as vector $\mathbf{a} \times \mathbf{b}$, but points in the opposite direction. Both cross products are perpendicular to the plane containing the two original vectors.

UNIT VECTORS

Any vector **a**, reduced to standard form so its starting point is at the origin, ends up at some point (x_a, y_a, z_a). This vector can be broken down into the sum of three mutually perpendicular vectors, each of which lies along one of the coordinate axes as shown in Fig. 9-9:

$$\begin{aligned}\mathbf{a} &= (x_a, y_a, z_a)\\ &= (x_a, 0, 0) + (0, y_a, 0) + (0, 0, z_a)\\ &= x_a(1,0,0) + y_a(0,1,0) + z_a(0,0,1)\end{aligned}$$

The vectors (1,0,0), (0,1,0), and (0,0,1) are called *unit vectors* because their length is 1. It is customary to name these vectors **i**, **j**, and **k**, because they come in handy:

$$\begin{aligned}(1,0,0) &= \mathbf{i}\\ (0,1,0) &= \mathbf{j}\\ (0,0,1) &= \mathbf{k}\end{aligned}$$

Therefore, the vector **a** shown in Fig. 9-9 breaks down this way:

$$\mathbf{a} = (x_a, y_a, z_a) = x_a\mathbf{i} + y_a\mathbf{j} + z_a\mathbf{k}$$

PROBLEM 9-6

Break the vector $\mathbf{b} = (-2,3,-7)$ down into a sum of multiples of the unit vectors **i**, **j**, and **k**.

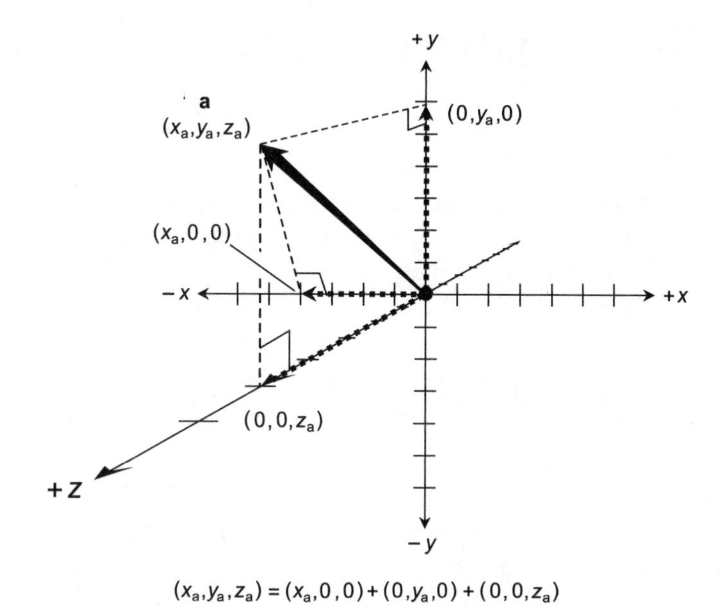

Fig. 9-9. Any vector in Cartesian three-space can be broken up into a sum of three component vectors, each of which lies on one of the coordinate axes.

SOLUTION 9-6

This is a simple process—almost trivial—but envisioning it requires a keen "mind's eye." If you have any trouble seeing this in your imagination, think of **i** as "one unit of width going to the right," **j** as "one unit of height going up," and **k** as "one unit of depth coming towards you." Here we go:

$$\begin{aligned}\mathbf{b} &= (-2,3,-7)\\ &= -2\times(1,0,0) + 3\times(0,1,0) + [-7\times(0,0,1)]\\ &= -2\mathbf{i} + 3\mathbf{j} + (-7)\mathbf{k}\\ &= -2\mathbf{i} + 3\mathbf{j} - 7\mathbf{k}\end{aligned}$$

Planes

The equation of a flat geometric plane in Cartesian 3D coordinates is somewhat like the equation of a straight line in Cartesian 2D coordinates.

CRITERIA FOR UNIQUENESS

A geometric plane in space can be uniquely defined according to any of the following criteria:

- Three points that do not all lie on the same straight line
- A point in the plane and a vector perpendicular to the plane
- Two intersecting straight lines
- Two parallel straight lines

GENERAL EQUATION OF PLANE

The simplest equation for a plane is derived on the basis of the second of the foregoing criteria: a point in the plane and a vector normal (perpendicular) to the plane. Figure 9-10 shows a plane W in Cartesian three-space, a point $P = (x_0,y_0,z_0)$ in plane W, and a vector $(a,b,c) = a\mathbf{i} + b\mathbf{j} + c\mathbf{k}$ that is normal to plane W. The vector (a,b,c) in this illustration is shown originating at point P, and not at the origin, because this particular plane does not contain the origin (0,0,0). The values $x = a$, $y = b$, and $z = c$ for the vector are nevertheless based on its standard form.

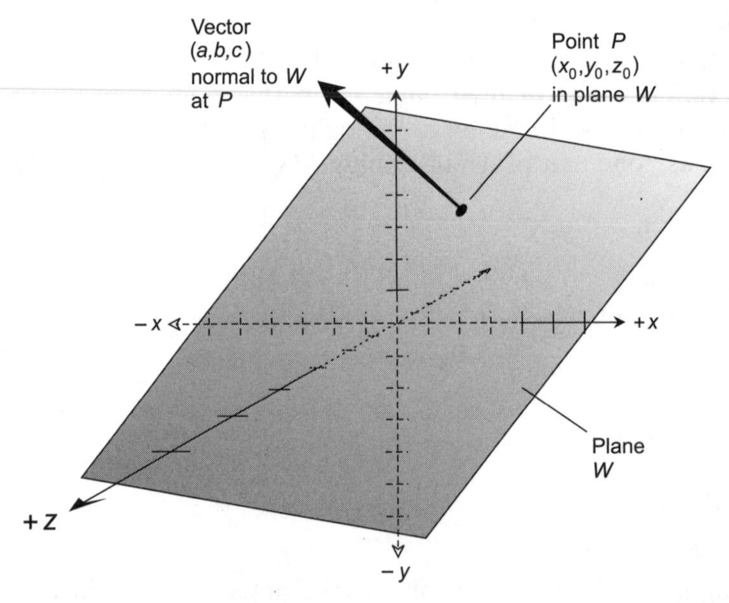

Fig. 9-10. A plane W can be uniquely defined on the basis of a point P in the plane and a vector (a,b,c) normal to the plane. Dashed portions of the coordinate axes are "behind" the plane.

When these things about a plane are known, we have enough information to uniquely define it and write its equation as follows:

$$a(x - x_0) + b(y - y_0) + c(z - z_0) = 0$$

In this form of the equation for a plane, the constants a, b, and c are called the *coefficients*. The above equation can also be written in this form:

$$ax + by + cz + d = 0$$

where

$$d = -(ax_0 + by_0 + cz_0) = -ax_0 - by_0 - cz_0$$

PLOTTING A PLANE

In order to draw a graph of a plane based on its equation, it is sufficient to obtain the points where the plane crosses each of the three coordinate axes. The plane can then be visualized, based on these points.

Not all planes cross all three of the axes in Cartesian xyz-space. If a plane is parallel to one of the axes, it does not cross that axis; it may cross one or both of the others. If a plane is parallel to the plane formed by two of the three axes, then it crosses only the axis to which it is not parallel. Any plane in Cartesian three-space must, however, cross at least one of the coordinate axes at some point.

PROBLEM 9-7

Draw a graph of the plane W represented by the following equation:

$$-2x - 4y + 3z - 12 = 0$$

SOLUTION 9-7

The x-intercept, or the point where the plane W intersects the x axis, can be found by setting $y = 0$ and $z = 0$ and solving the resulting equation for x. Call this point P:

$$-2x - 4 \times 0 + 3 \times 0 - 12 = 0$$
$$-2x - 12 = 0$$
$$-2x = 12$$
$$x = 12/(-2) = -6$$

Therefore,

$$P = (-6,0,0)$$

The y-intercept, or the point where the plane W intersects the y axis, can be found by setting $x = 0$ and $z = 0$ and solving the resulting equation for y. Call this point Q:

$$-2 \times 0 - 4y + 3 \times 0 - 12 = 0$$
$$-4y - 12 = 0$$
$$-4y = 12$$
$$y = 12/(-4) = -3$$

Therefore,

$$Q = (0, -3,0)$$

The z-intercept, or the point where the plane W intersects the z axis, can be found by setting $x = 0$ and $y = 0$ and solving the resulting equation for z. Call this point R:

$$-2 \times 0 - 4 \times 0 + 3z - 12 = 0$$
$$3z - 12 = 0$$
$$3z = 12$$
$$z = 12/3 = 4$$

Therefore,

$$R = (0,0,4)$$

These three points are shown in the plot of Fig. 9-11. The plane can be envisioned, based on this data. (The dashed axes are "behind" the plane.)

PROBLEM 9-8

Suppose a plane contains the point (2,–7,0) and a normal vector to the plane at this point is 3**i** + 3**j** + 2**k**. What is the equation of this plane?

SOLUTION 9-8

The vector 3**i** + 3**j** + 2**k** is equivalent to $(a,b,c) = (3,3,2)$. We have one point $(x_0,y_0,z_0) = (2,-7,0)$. Plugging these values into the general formula for the equation of a plane gives us the following:

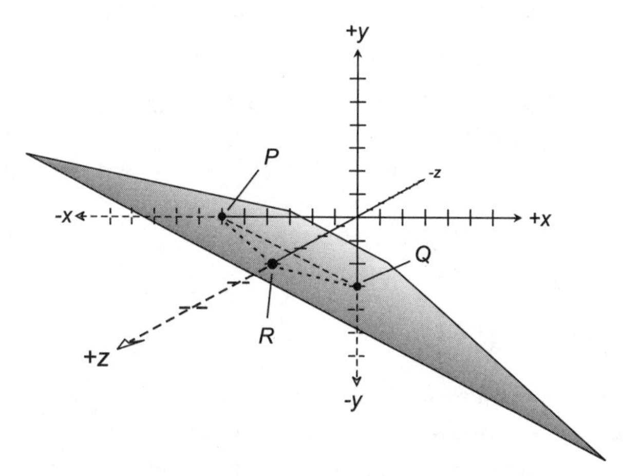

Fig. 9-11. Illustration for Problem 9-7. Dashed portions of the coordinate axes are "behind" the plane.

$$a(x - x_0) + b(y - y_0) + c(z - z_0) = 0$$
$$3(x - 2) + 3[y - (-7)] + 2(z - 0) = 0$$
$$3(x - 2) + 3(y + 7) + 2z = 0$$
$$3x - 6 + 3y + 21 + 2z = 0$$
$$3x + 3y + 2z + 15 = 0$$

Straight Lines

Straight lines in Cartesian three-space present a more complicated picture than straight lines in the Cartesian coordinate plane. This is because there is an added dimension, making the expression of the direction more complex. But all linear equations, no matter what the number of dimensions, have one thing in common: they can be reduced to a form where no variable is raised to any power other than 0 or 1.

SYMMETRIC-FORM EQUATION

A straight line in Cartesian three-space can be represented by a "three-way" equation in three variables. This equation is known as a *symmetric-form equation*. It takes the following form, where x, y, and z are the variables, (x_0,y_0,z_0) represents the coordinates of a specific point on the line, and a, b, and c are constants:

$$(x - x_0)/a = (y - y_0)/b = (z - z_0)/c$$

This requires that none of the three constants a, b, or c be equal to zero. If $a = 0$ or $b = 0$ or $c = 0$, the result is a zero denominator in one of the expressions, and division by zero is not defined.

DIRECTION NUMBERS

In the symmetric-form equation of a straight line, the constants a, b, and c are known as the *direction numbers*. If we consider a vector **m** with its end point at the origin and its "arrowed end" at the point $(x,y,z) = (a,b,c)$, then the vector **m** is parallel to the line denoted by the symmetric-form equation. We have:

$$\mathbf{m} = a\mathbf{i} + b\mathbf{j} + c\mathbf{k}$$

where **m** is the three-dimensional equivalent of the slope of a line in the Cartesian plane. This is shown in Fig. 9-12 for a line L containing a point $P = (x_0,y_0,z_0)$.

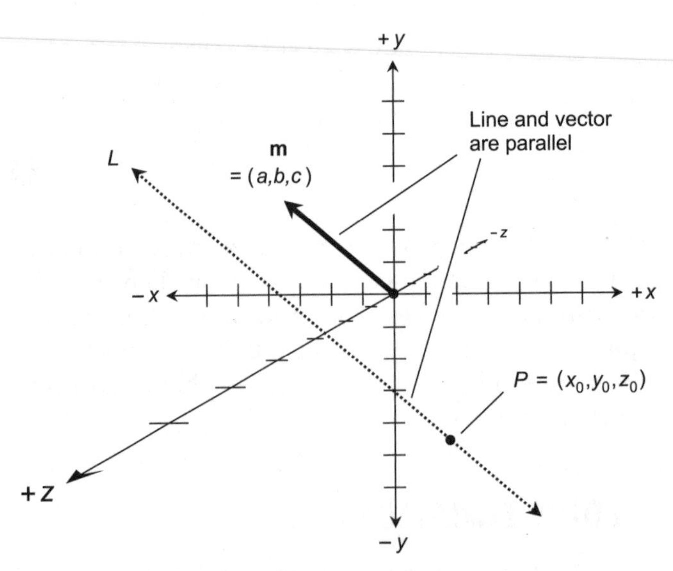

Fig. 9-12. A line L can be uniquely defined on the basis of a point P on the line and a vector $\mathbf{m} = (a,b,c)$ that is parallel to the line.

PARAMETRIC EQUATIONS

There are infinitely many vectors that can satisfy the requirement for **m**. If t is any nonzero real number, then $t\mathbf{m} = (ta,tb,tc) = ta\mathbf{i} + tb\mathbf{j} + tc\mathbf{k}$ will work just as well as **m** for the purpose of defining the direction of a line L. This gives us an alternative form for the equation of a line in Cartesian three-space:

$$x = x_0 + at$$
$$y = y_0 + bt$$
$$z = z_0 + ct$$

The nonzero real number t is called a *parameter*, and the above set of equations is known as a set of *parametric equations* for a straight line in xyz-space. In order for an entire line (straight, and infinitely long) to be defined on this basis of parametric equations, the parameter t must be allowed to range over the entire set of real numbers, including zero.

PROBLEM 9-9

Find the symmetric-form equation for the line L shown in Fig. 9-13.

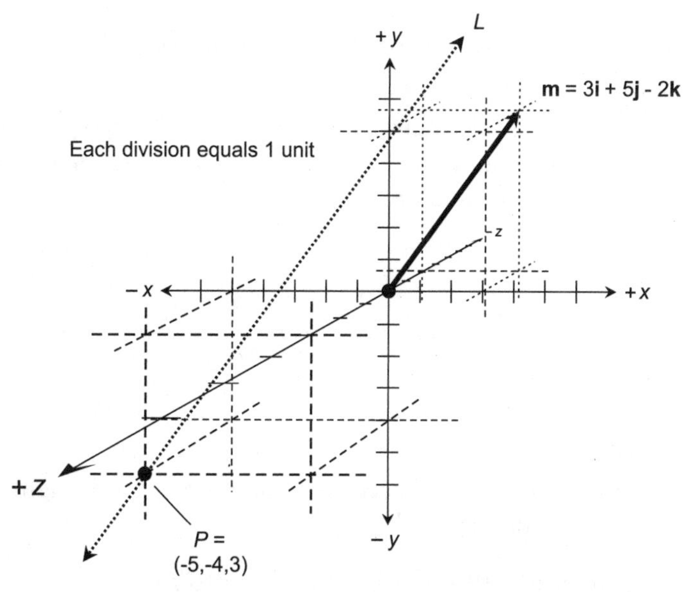

Fig. 9-13. Illustration for Problems 9-9 and 9-10.

SOLUTION 9-9

The line L passes through the point $P = (-5,-4,3)$ and is parallel to the vector $\mathbf{m} = 3\mathbf{i} + 5\mathbf{j} - 2\mathbf{k}$. The direction numbers of L are the coefficients of the vector **m**, that is:

$$a = 3$$
$$b = 5$$
$$c = -2$$

We are given a point P on L such that:

$$x_0 = -5$$
$$y_0 = -4$$
$$z_0 = 3$$

Plugging these values into the general symmetric-form equation for a line in Cartesian three-space gives us this:

$$(x - x_0)/a = (y - y_0)/b = (z - z_0)/c$$
$$[x - (-5)]/3 = [y - (-4)]/5 = (z - 3)/(-2)$$
$$(x + 5)/3 = (y + 4)/5 = (z - 3)/(-2)$$

PROBLEM 9-10

Find a set of parametric equations for the line L shown in Fig. 9-13.

SOLUTION 9-10

This involves nothing more than rearranging the values of x_0, y_0, z_0, a, b, and c in the symmetric-form equation, and rewriting the data in the form of parametric equations. The results are:

$$x = -5 + 3t$$
$$y = -4 + 5t$$
$$z = 3 - 2t$$

Quiz

Refer to the text in this chapter if necessary. A good score is eight correct. Answers are in the back of the book.

1. The dot product (3,5,0) • (–4,–6,2) is equal to
   (a) the scalar quantity –4
   (b) the vector (–12,–30,0)

(c) the scalar quantity –42
(d) a vector perpendicular to the plane containing them both

2. What does the graph of the equation $y = 3$ look like in Cartesian three-space?
(a) A plane perpendicular to the y axis
(b) A plane parallel to the xy plane
(c) A line parallel to the y axis
(d) A line parallel to the xy plane

3. Suppose vector **d** in the Cartesian plane begins at exactly (1,1) and ends at exactly (4,0). What is dir **d**, expressed to the nearest degree?
(a) 342°
(b) 18°
(c) 0°
(d) 90°

4. Suppose a line is represented by the equation $(x - 3)/2 = (y + 4)/5 = z - 1$. Which of the following is a point on this line?
(a) (–3,4,–1)
(b) (3,–4,1)
(c) (2,5,1)
(d) There is no way to determine such a point without more information

5. Let $\triangle PQR$ be a right triangle whose hypotenuse measures exactly 1 unit in length, and whose other two sides measure p meters and q meters, as shown in Fig. 9-14. Let θ be the angle whose apex is point P. Which of the following statements is true?
(a) $\sin \theta = p/q$
(b) $\sin \theta = \cos \phi$
(c) $\tan \phi = \tan \theta$
(d) $\cos \theta = 1/(\tan \phi)$

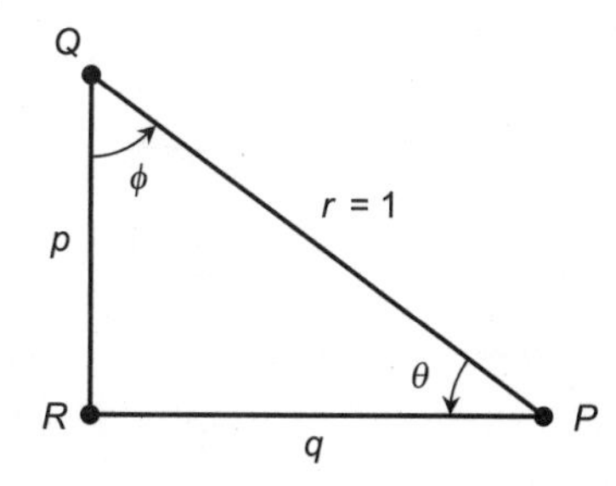

Fig. 9-14. Illustration for quiz questions 5 and 6.

6. With reference to Fig. 9-14, the Pythagorean theorem can be used to demonstrate that
   (a) $\sin\theta + \cos\phi = 1$
   (b) $\tan\theta - \tan\phi = 1$
   (c) $(\sin\theta)^2 + (\cos\theta)^2 = 1$
   (d) $(\sin\theta)^2 - (\cos\theta)^2 = 1$

7. What is the cross product $(2\mathbf{i} + 0\mathbf{j} + 0\mathbf{k}) \times (0\mathbf{i} + 2\mathbf{j} + 0\mathbf{k})$?
   (a) $0\mathbf{i} + 0\mathbf{j} + 0\mathbf{k}$
   (b) $2\mathbf{i} + 2\mathbf{j} + 0\mathbf{k}$
   (c) $0\mathbf{i} + 0\mathbf{j} + 4\mathbf{k}$
   (d) The scalar 0

8. What is the sum of the two vectors (3,5) and (–5,–3)?
   (a) (0,0)
   (b) (8,8)
   (c) (2,2)
   (d) (–2,2)

9. If a straight line in Cartesian three-space has direction defined by $\mathbf{m} = 0\mathbf{i} + 0\mathbf{j} + 3\mathbf{k}$, we can surmise
   (a) that the line is parallel to the x axis
   (b) that the line lies in the yz plane
   (c) that the line lies in the xy plane
   (d) none of the above

10. Suppose a plane passes through the origin, and a vector normal to the plane is represented by $4\mathbf{i} - 5\mathbf{j} + 8\mathbf{k}$. The equation of this plane is
    (a) $4x - 5y + 8z = 0$
    (b) $-4x + 5y - 8z = 7$
    (c) $(x - 4) = (y + 5) = (z - 8)$
    (d) impossible to determine without more information

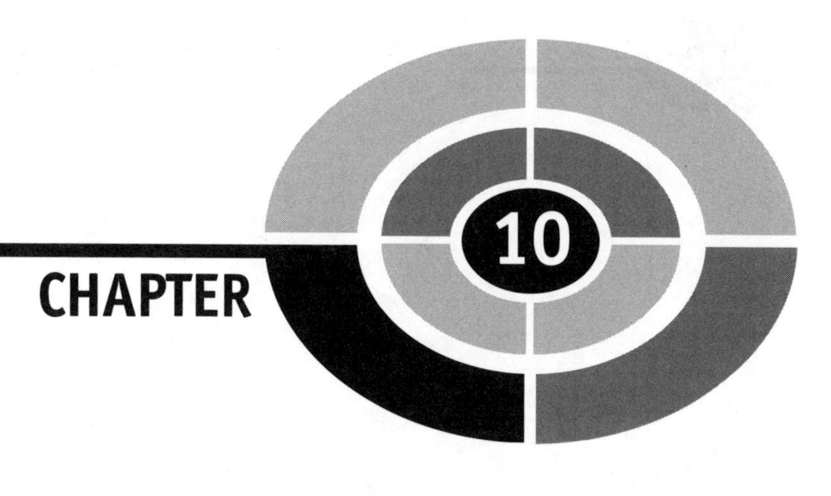

Alternative Coordinates

Cartesian coordinates do not represent the only way that points can be located on a plane or in 3D space. In this chapter we'll look at polar, latitude/longitude, cylindrical, and spherical schemes.

Polar Coordinates

Two versions of the *polar coordinate plane* are shown in Figs. 10-1 and 10-2. The independent variable is plotted as an angle θ relative to a reference axis pointing to the right (or "east"), and the dependent variable is plotted as a distance (called the *radius*) r from the origin. A coordinate point is thus denoted in the form of an ordered pair (θ,r).

THE RADIUS

In any polar plane, the radii are shown by concentric circles. The larger the circle, the greater the value of r. In Figs. 10-1 and 10-2, the circles are not

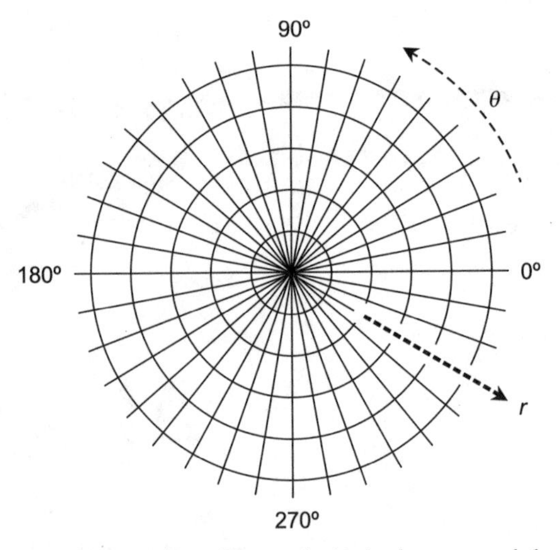

Fig. 10-1. The polar coordinate plane. The angle θ is in degrees, and the radius r is in uniform increments.

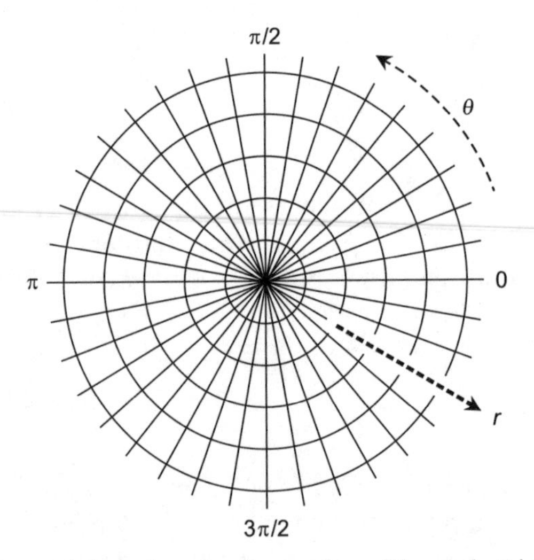

Fig. 10-2. Another form of the polar coordinate plane. The angle θ is in radians, and the radius r is in uniform increments.

labeled in units. You can do that for yourself. Imagine each concentric circle, working outward, as increasing by any number of units you want. For example, each radial division might represent one unit, or five units, or 10, or 100.

THE DIRECTION

Direction can be expressed in degrees or radians counterclockwise from a reference axis pointing to the right or "east." In Fig. 10-1, the direction θ is in degrees. Figure 10-2 shows the same polar plane, using radians to express the direction. (The "rad" abbreviation is not used, because it is obvious from the fact that the angles are multiples of π.) Regardless of whether degrees or radians are used, the angular scale is linear. The physical angle on the graph is directly proportional to the value of θ.

NEGATIVE RADII

In polar coordinates, it is all right to have a negative radius. If some point is specified with $r < 0$, we multiply r by -1 so it becomes positive, and then add or subtract 180° (π rad) to or from the direction. That's like saying, "Go 10 kilometers east" instead of "Go minus 10 kilometers west." Negative radii must be allowed in order to graph figures that represent functions whose ranges can attain negative values.

NON-STANDARD DIRECTIONS

It's okay to have non-standard direction angles in polar coordinates. If the value of θ is 360° (2π rad) or more, it represents more than one complete counterclockwise revolution from the 0° (0 rad) reference axis. If the direction angle is less than 0° (0 rad), it represents clockwise revolution instead of counterclockwise revolution. Non-standard direction angles must be allowed in order to graph figures that represent functions whose domains go outside the standard angle range.

PROBLEM 10-1

Provide an example of a graphical object that can be represented as a function in polar coordinates, but not in Cartesian coordinates.

SOLUTION 10-1

Recall the definitions of the terms *relation* and *function* from Chapter 6. When we talk about a function f, we can say that $r = f(\theta)$. A simple function of θ in polar coordinates is a *constant function* such as this:

$$f(\theta) = 3$$

Because $f(\theta)$ is just another way of denoting r, the radius, this function tells us that $r = 3$. This is a circle with a radius of 3 units.

In Cartesian coordinates, the equation of the circle with radius of 3 units is more complicated. It looks like this:

$$x^2 + y^2 = 9$$

(Note that $9 = 3^2$, the square of the radius.) If we let y be the dependent variable and x be the independent variable, we can rearrange the equation of the circle to get:

$$y = \pm(9 - x^2)^{1/2}$$

If we say that $y = g(x)$ where g is a function of x in this case, we are mistaken. There are values of x (the independent variable) that produce two values of y (the dependent variable). For example, when $x = 0$, $y = \pm 3$. If we want to say that g is a relation, that's fine, but we cannot call it a function.

Some Examples

In order to get a good idea of how the polar coordinate system works, let's look at the graphs of some familiar objects. Circles, ellipses, spirals, and other figures whose equations are complicated in Cartesian coordinates can often be expressed much more simply in polar coordinates. In general, the polar direction θ is expressed in radians. In the examples that follow, the "rad" abbreviation is eliminated, because it is understood that all angles are in radians.

CIRCLE CENTERED AT ORIGIN

The equation of a *circle* centered at the origin in the polar plane is given by the following formula:

$$r = a$$

where a is a real-number constant greater than 0. This is illustrated in Fig. 10-3.

CIRCLE PASSING THROUGH ORIGIN

The general form for the equation of a circle passing through the origin and centered at the point (θ_0, r_0) in the polar plane (Fig. 10-4) is as follows:

$$r = 2r_0 \cos(\theta - \theta_0)$$

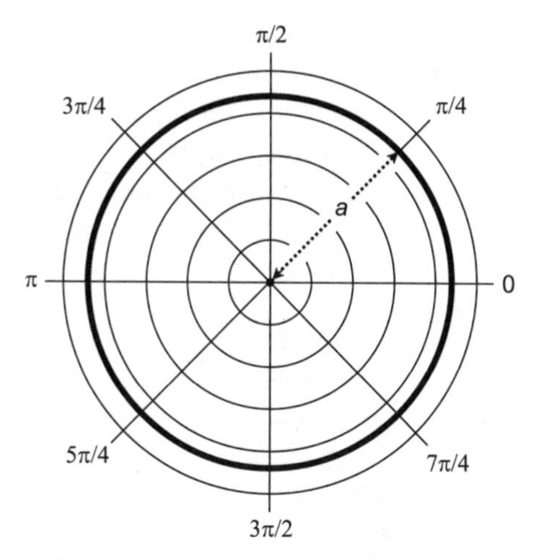

Fig. 10-3. Polar graph of a circle centered at the origin.

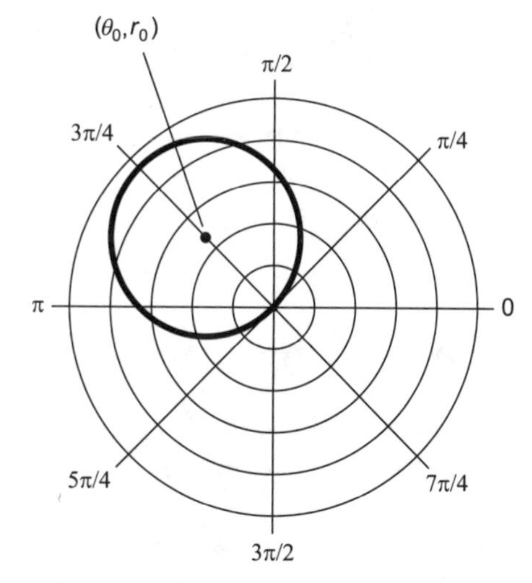

Fig. 10-4. Polar graph of a circle passing through the origin.

ELLIPSE CENTERED AT ORIGIN

The equation of an *ellipse* centered at the origin in the polar plane is given by the following formula:

$$r = ab/(a^2 \sin^2 \theta + b^2 \cos^2 \theta)^{1/2}$$

where a and b are real-number constants greater than 0.

In the ellipse, a represents the distance from the origin to the curve as measured along the "horizontal" ray $\theta = 0$, and b represents the distance from the origin to the curve as measured along the "vertical" ray $\theta = \pi/2$. This is illustrated in Fig. 10-5. The values $2a$ and $2b$ represent the lengths of the *semi-axes* of the ellipse; the greater value is the length of the *major semi-axis*, and the lesser value is the length of the *minor semi-axis*.

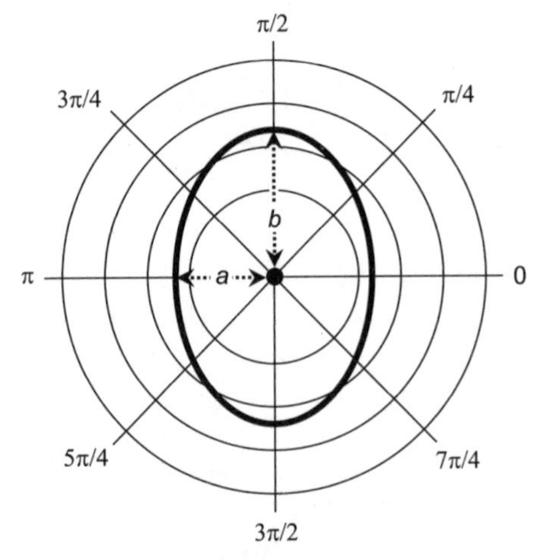

Fig. 10-5. Polar graph of an ellipse centered at the origin.

HYPERBOLA CENTERED AT ORIGIN

The general form of the equation of a *hyperbola* centered at the origin in the polar plane is given by the following formula:

$$r = ab/(a^2 \sin^2 \theta - b^2 \cos^2 \theta)^{1/2}$$

where a and b are real-number constants greater than 0.

Let D represent a rectangle whose center is at the origin, whose vertical edges are tangent to the hyperbola, and whose vertices (corners) lie on the *asymptotes* of the hyperbola (Fig. 10-6). Let a represent the distance from the origin to D as measured along the "horizontal" ray $\theta = 0$, and let b represent the distance from the origin to D as measured along the "vertical" ray $\theta = \pi/2$. The values $2a$ and $2b$ represent the lengths of the *semi-axes* of the hyperbola; the greater value is the length of the *major semi-axis*, and the lesser value is the length of the *minor semi-axis*.

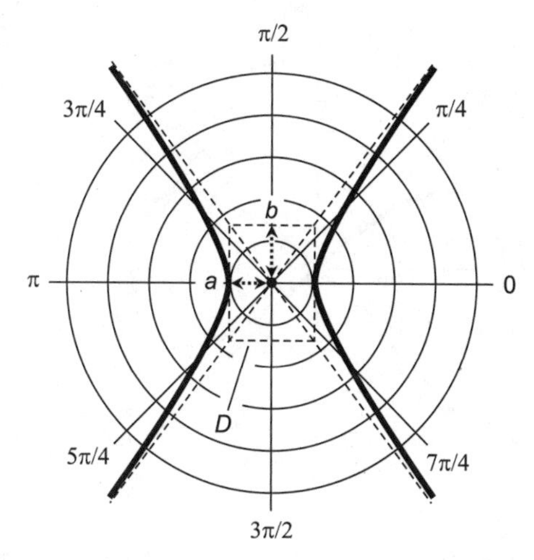

Fig. 10-6. Polar graph of a hyperbola centered at the origin.

LEMNISCATE

The general form of the equation of a *lemniscate* centered at the origin in the polar plane is given by the following formula:

$$r = a\,(\cos 2\theta)^{1/2}$$

where a is a real-number constant greater than 0. This is illustrated in Fig. 10-7. The area A of each loop of the figure is given by:

$$A = a^2$$

THREE-LEAFED ROSE

The general form of the equation of a *three-leafed rose* centered at the origin in the polar plane is given by either of the following two formulas:

$$r = a \cos 3\theta$$
$$r = a \sin 3\theta$$

where a is a real-number constant greater than 0. The cosine version of the curve is illustrated in Fig. 10-8A. The sine version is illustrated in Fig. 10-8B.

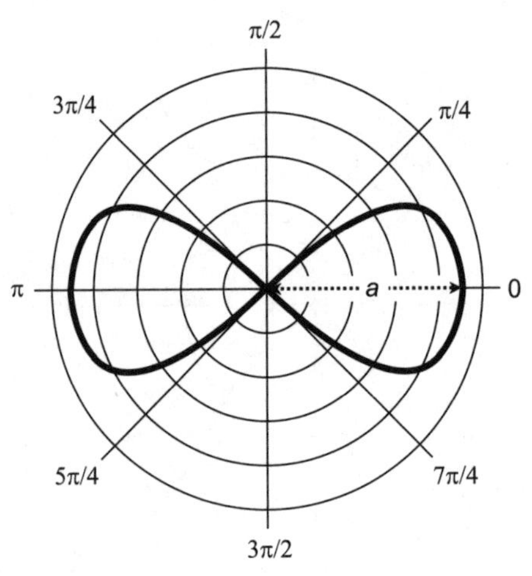

Fig. 10-7. Polar graph of a lemniscate centered at the origin.

FOUR-LEAFED ROSE

The general form of the equation of a *four-leafed rose* centered at the origin in the polar plane is given by either of the following two formulas:

$$r = a \cos 2\theta$$

$$r = a \sin 2\theta$$

where a is a real-number constant greater than 0. The cosine version is illustrated in Fig. 10-9A. The sine version is illustrated in Fig. 10-9B.

SPIRAL

The general form of the equation of a *spiral* centered at the origin in the polar plane is given by the following formula:

$$r = a\theta$$

where a is a real-number constant greater than 0. An example of this type of spiral, called the *spiral of Archimedes* because of the uniform manner in which its radius increases as the angle increases, is illustrated in Fig. 10-10.

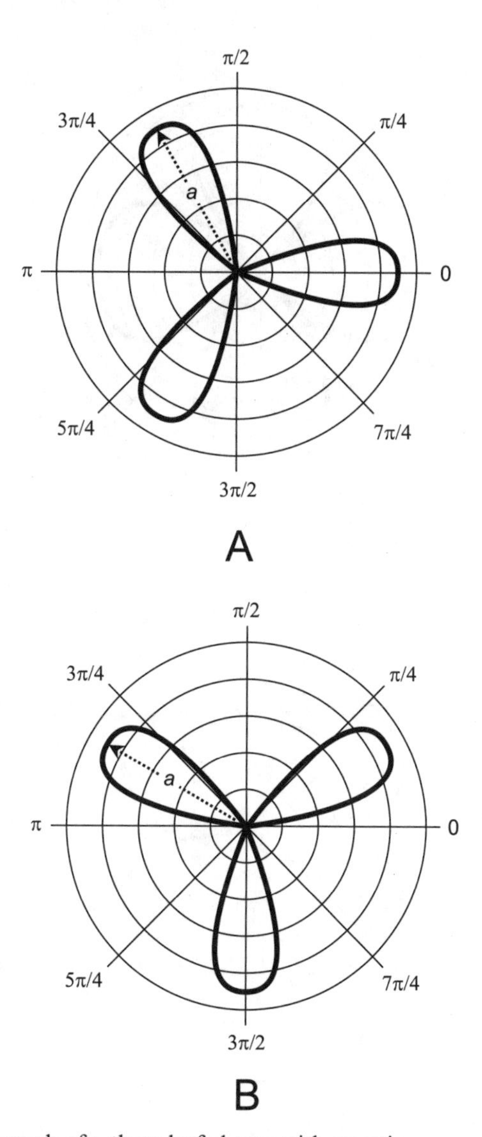

Fig. 10-8. (A) Polar graph of a three-leafed rose with equation $r = a \cos 3\theta$. (B) Polar graph of a three-leafed rose with equation $r = a \sin 3\theta$.

CARDIOID

The general form of the equation of a *cardioid* centered at the origin in the polar plane is given by the following formula:

$$r = 2a(1 + \cos \theta)$$

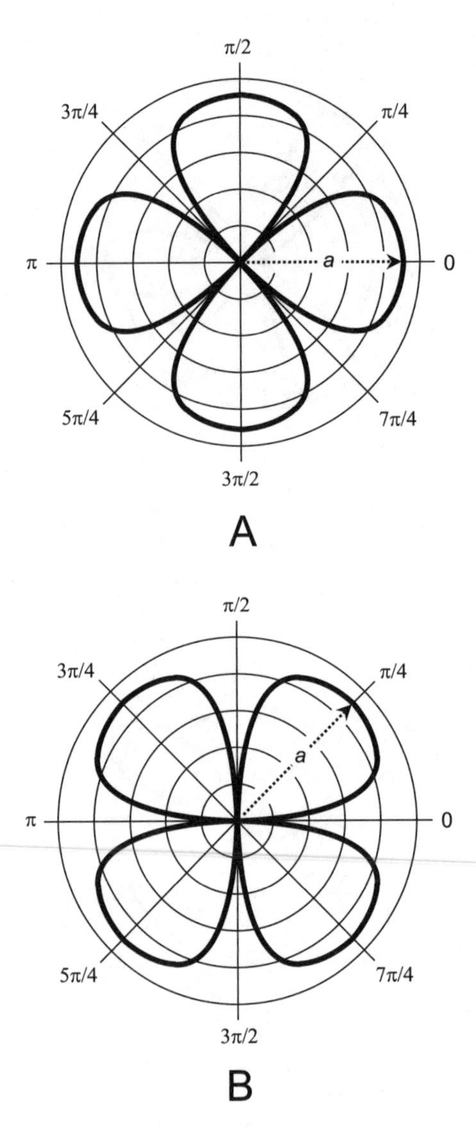

Fig. 10-9. (A) Polar graph of a four-leafed rose with equation $r = a \cos 2\theta$. (B) Polar graph of a four-leafed rose with equation $r = a \sin 2\theta$.

where a is a real-number constant greater than 0. An example of this type of curve is illustrated in Fig. 10-11.

PROBLEM 10-2

What is the value of the constant, a, in the spiral shown in Fig. 10-10? What is the equation of this spiral? Assume that each radial division represents 1 unit.

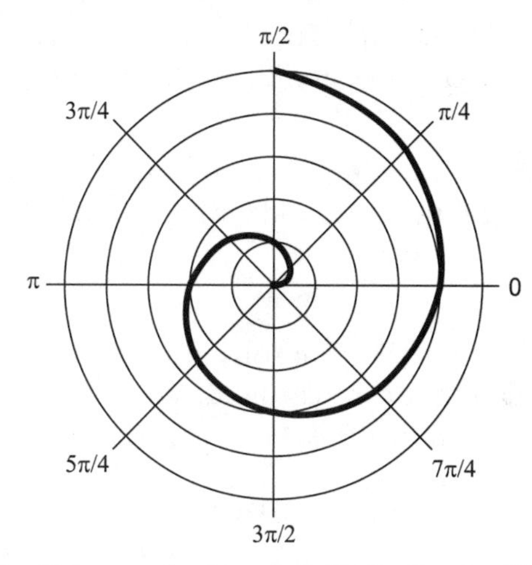

Fig. 10-10. Polar graph of a spiral; illustration for Problem 10-2.

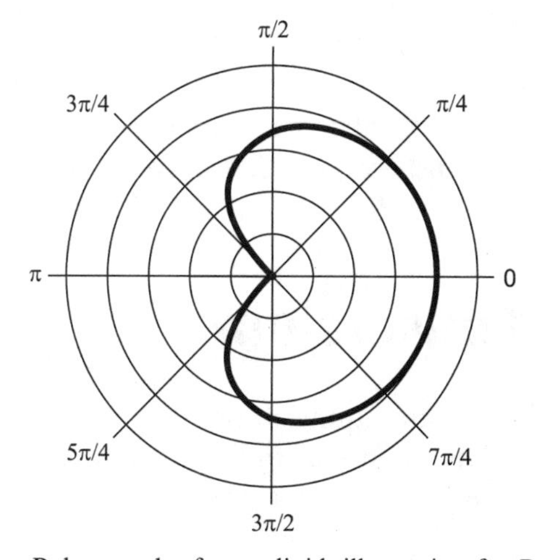

Fig. 10-11. Polar graph of a cardioid; illustration for Problem 10-3.

SOLUTION 10-2

Note that if $\theta = \pi$, then $r = 2$. Therefore, we can solve for a by substituting this number pair in the general equation for the spiral. We know that $(\theta,r) = (\pi,2)$, and that is all we need. Proceed like this:

$$r = a\theta$$

$$2 = a\pi$$

$$2/\pi = a$$

Therefore, $a = 2/\pi$, and the equation of the spiral is $r = (2/\pi)\theta$ or, in a somewhat simpler form without parentheses, $r = 2\theta/\pi$.

PROBLEM 10-3

What is the value of the constant, *a*, in the cardioid shown in Fig. 10-11? What is the equation of this cardioid? Assume that each radial division represents 1 unit.

SOLUTION 10-3

Note that if $\theta = 0$, then $r = 4$. We can solve for *a* by substituting this number pair in the general equation for the cardioid. We know that $(\theta,r) = (0,4)$, and that is all we need. Proceed like this:

$$r = 2a(1 + \cos\theta)$$
$$4 = 2a(1 + \cos 0)$$
$$4 = 2a(1 + 1)$$
$$4 = 4a$$
$$a = 1$$

This means that the equation of the cardioid is $r = 2(1 + \cos\theta)$ or, in a simpler form without parentheses, $r = 2 + 2\cos\theta$.

Compression and Conversion

Here are a couple of interesting things, one of which serves the imagination, and the other of which has extensive applications in science and engineering.

GEOMETRIC POLAR PLANE

Figure 10-12 shows a polar plane on which the radial scale is graduated geometrically. The point corresponding to 1 on the *r* axis is halfway between the origin and the outer periphery, which is labeled ∞ (the "infinity" symbol). Succeeding integer points are placed halfway between previous integer points and the outer periphery. In this way, the entire polar coordinate plane is depicted within a finite open circle.

The radial scale of this coordinate system can be expanded or compressed by multiplying all the values on the *r* axis by a constant. This allows various relations and functions to be plotted, minimizing distortion in particular

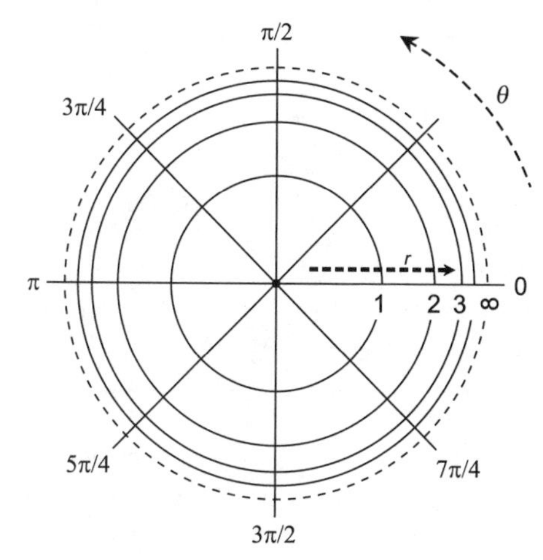

Fig. 10-12. A polar coordinate plane with a "geometrically compressed" radial axis.

regions of interest. Distortion relative to the conventional polar coordinate plane is greatest near the periphery, and is least near the origin.

This "geometric axis compression" scheme can also be used with the axes of rectangular coordinates in two or three dimensions. It is not often seen in the literature, but it can be interesting because it provides a "view to infinity" that other coordinate systems do not.

POLAR VS CARTESIAN CONVERSIONS

Figure 10-13 shows a point $P = (x_0, y_0) = (\theta_0, r_0)$ graphed on superimposed Cartesian and polar coordinate systems. If we know the Cartesian coordinates, we can convert to polar coordinates using these formulas:

$$\theta_0 = \arctan (y_0/x_0) \text{ if } x_0 > 0$$
$$\theta_0 = 180^\circ + \arctan (y_0/x_0) \text{ if } x_0 < 0 \text{ (for } \theta_0 \text{ in degrees)}$$
$$\theta_0 = \pi + \arctan (y_0/x_0) \text{ if } x_0 < 0 \text{ (for } \theta_0 \text{ in radians)}$$
$$r_0 = (x_0^2 + y_0^2)^{1/2}$$

We can't have $x_0 = 0$ because that produces an undefined quotient. If a value of θ_0 thus determined happens to be negative, add 360° or 2π rad to get the "legitimate" value.

Polar coordinates are converted to Cartesian coordinates by the following formulas:

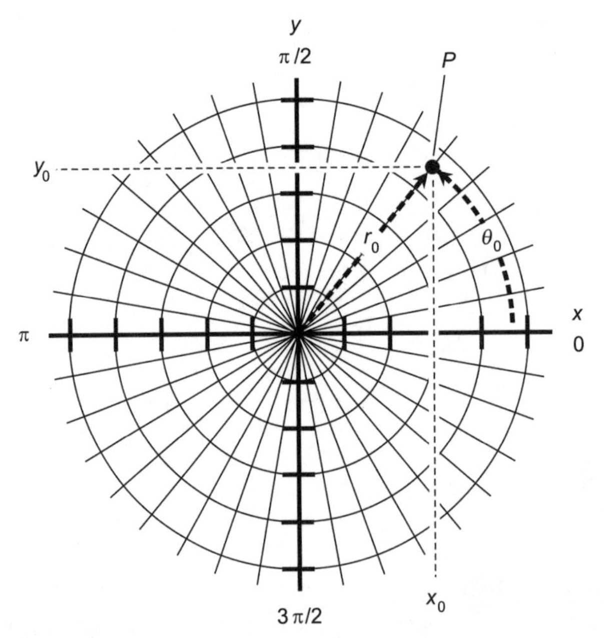

Fig. 10-13. Conversion between polar and Cartesian (rectangular) coordinates. Each radial division represents one unit. Each division on the x and y axes also represents one unit.

$$x_0 = r_0 \cos \theta_0$$
$$y_0 = r_0 \sin \theta_0$$

These same formulas can be used, by means of substitution, to convert Cartesian-coordinate relations to polar-coordinate relations, and vice versa. The generalized Cartesian-to-polar conversion formulas look like this:

$$\theta = \arctan (y/x) \text{ if } x > 0$$
$$\theta = 180^\circ + \arctan (y/x) \text{ if } x < 0 \text{ (for } \theta \text{ in degrees)}$$
$$\theta = \pi + \arctan (y/x) \text{ if } x < 0 \text{ (for } \theta \text{ in radians)}$$
$$r = (x^2 + y^2)^{1/2}$$

The generalized polar-to-Cartesian conversion formulas are:

$$x = r \cos \theta$$
$$y = r \sin \theta$$

When making a conversion from polar to Cartesian coordinates or vice versa, a relation that is a function in one system may be a function in the other, but this is not always true.

PROBLEM 10-4
Consider the point $(\theta_0, r_0) = (135°, 2)$ in polar coordinates. What is the (x_0, y_0) representation of this point in Cartesian coordinates?

SOLUTION 10-4
Use the conversion formulas above:

$$x_0 = r_0 \cos \theta_0$$
$$y_0 = r_0 \sin \theta_0$$

Plugging in the numbers gives us these values, accurate to three decimal places:

$$x_0 = 2 \cos 135° = 2 \times (-0.707) = -1.414$$
$$y_0 = 2 \sin 135° = 2 \times 0.707 = 1.414$$

Thus, $(x_0, y_0) = (-1.414, 1.414)$.

The Navigator's Way

Navigators and military people use a form of coordinate plane similar to that used by mathematicians. The radius is more often called the *range*, and real-world units are commonly specified, such as meters (m) or kilometers (km). The angle, or direction, is more often called the *azimuth*, *heading*, or *bearing*, and is measured in degrees clockwise from north. The basic scheme is shown in Fig. 10-14. The azimuth is symbolized α (the lowercase Greek alpha), and the range is symbolized r.

WHAT IS NORTH?

There are two ways of defining "north," or 0°. The more accurate, and thus the preferred and generally accepted, standard uses *geographic north*. This is the direction you would travel if you wanted to take the shortest possible route over the earth's surface to the north geographic pole. The less accurate standard uses *magnetic north*. This is the direction indicated by the needle in a magnetic compass.

For most locations on the earth's surface, there is some difference between geographic north and magnetic north. This difference, measured in degrees, is called the *declination*. Navigators in olden times had to know the declination for their location, when they couldn't use the stars to determine geographic

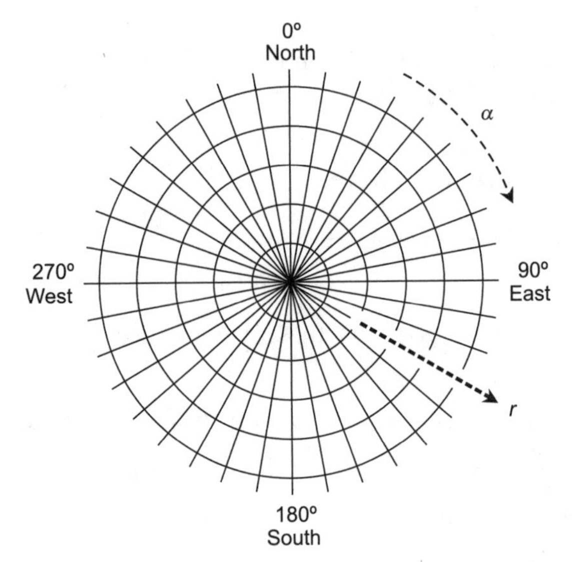

Fig. 10-14. The navigator's polar coordinate plane. The bearing α is in degrees, and the range r is in arbitrary units.

north. Nowadays, there are electronic navigation systems such as the *Global Positioning System* (GPS) that make the magnetic compass irrelevant, provided the equipment is in working order. (Most oceangoing vessels still have magnetic compasses on board, just in case of a failure of the more sophisticated equipment.)

STRICT RESTRICTIONS

In navigator's polar coordinates, the range can never be negative. No navigator ever talks about traveling –20 km on a heading of 270°, for example, when they really mean to say they are traveling 20 km on a heading of 90°. When working out certain problems, it's possible that the result might contain a negative range. If this happens, the value of r should be multiplied by -1 and the value of α should be increased or decreased by 180° so the result is at least 0° but less than 360°.

The azimuth, bearing, or heading must likewise conform to certain values. The smallest possible value of α is 0° (representing geographic north). As you turn clockwise as seen from above, the values of α increase through 90° (east), 180° (south), 270° (west), and ultimately approach, but never reach, 360° (north again).

We therefore have these restrictions on the ordered pair (α,r):

$$0^\circ \leq \alpha < 360^\circ$$
$$r \geq 0$$

MATHEMATICIAN'S POLAR VS NAVIGATOR'S POLAR

Sometimes it is necessary to convert from mathematician's polar coordinates (let's call them MPC for short) to navigator's polar coordinates (NPC), or vice versa. When making the conversion, the radius of a particular point, r_0, is the same in both systems, so no change is necessary. But the angles differ.

If you know the direction angle θ_0 of a point in MPC and you want to find the equivalent azimuth α_0 in NPC, first be sure θ_0 is expressed in degrees, not radians. Then you can use either of the following conversion formulas, depending on the value of θ_0:

$$\alpha_0 = 90^\circ - \theta_0 \text{ if } 0^\circ \leq \theta_0 \leq 90^\circ$$
$$\alpha_0 = 450^\circ - \theta_0 \text{ if } 90^\circ < \theta_0 < 360^\circ$$

If you know the azimuth α_0 of a distant point in NPC and you want to find the equivalent direction angle θ_0 in MPC, then you can use either of the following conversion formulas, depending on the value of α_0:

$$\theta_0 = 90^\circ - \alpha_0 \text{ if } 0^\circ \leq \alpha_0 \leq 90^\circ$$
$$\theta_0 = 450^\circ - \alpha_0 \text{ if } 90^\circ < \alpha_0 < 360^\circ$$

NAVIGATOR'S POLAR VS CARTESIAN

Now suppose that you want to convert from NPC to Cartesian coordinates. Here are the conversion formulas for translating the coordinates for a point (α_0,r_0) in NPC to a point (x_0,y_0) in the Cartesian plane:

$$x_0 = r_0 \sin \alpha_0$$
$$y_0 = r_0 \cos \alpha_0$$

These are similar to the formulas you use to convert MPC to Cartesian coordinates, except that the roles of the sine and cosine function are reversed.

In order to convert the coordinates of a point (x_0,y_0) in Cartesian coordinates to a point (α_0,r_0) in NPC, use these formulas:

$$\alpha_0 = \arctan\,(x_0/y_0) \text{ if } y_0 > 0$$
$$\alpha_0 = 180^\circ + \arctan\,(x_0/y_0) \text{ if } y_0 < 0$$
$$r_0 = (x_0^2 + y_0^2)^{1/2}$$

We can't have $y_0 = 0$, because that produces an undefined quotient. If a value of α_0 thus determined happens to be negative, add 360° to get the "legitimate" value. These are similar to the formulas for converting Cartesian coordinates to MPC.

PROBLEM 10-5

Suppose a radar set, that uses NPC, indicates the presence of a hovering object at a bearing of 300° and a range of 40 km. If we say that a kilometer is the same as a "unit," what are the coordinates (θ_0,r_0) of this object in mathematician's polar coordinates? Express θ_0 in both degrees and radians.

SOLUTION 10-5

We are given coordinates $(\alpha_0,r_0) = (300°,40)$. The value of r_0, the radius, is the same as the range, in this case 40 units. As for the angle θ_0, remember the conversion formulas given above. In this case, because α_0 is greater than 90° and less than 360°:

$$\theta_0 = 450^\circ - \alpha_0$$
$$= 450^\circ - 300^\circ = 150^\circ$$

Therefore, $(\theta_0,r_0) = (150°,40)$. To express θ_0 in radians, recall that there are 2π radians in a full 360° circle, or π radians in a 180° angle. Note that 150° is exactly 5/6 of 180°. Therefore, $\theta_0 = 5\pi/6$ rad, and we can say that $(\theta_0,r_0) = (150°,40) = (5\pi/6,40)$. We can leave the "rad" off the angle designator here. When units are not specified for an angle, radians are assumed.

PROBLEM 10-6

Suppose you are on an archeological expedition, and you unearth a stone on which appears a treasure map. The map says "You are here" next to an X, and then says, "Go north 40 paces and then west 30 paces." Suppose that you let west represent the negative x axis of a Cartesian coordinate system, east represent the positive x axis, south represent the negative y axis, and north represent the positive y axis. Also suppose that you let one "pace" represent one "unit" of radius, and also one "unit" in the Cartesian system. If you are naïve enough to look for the treasure and lazy enough so you insist on walking in a straight line to reach it, how many paces should you travel, and in what direction, in navigator's polar coordinates? Determine your answer to the nearest degree, and to the nearest pace.

SOLUTION 10-6

Determine the ordered pair in Cartesian coordinates that corresponds to the imagined treasure site. Consider the origin to be the spot where the map was unearthed. If we let (x_0,y_0) be the point where the treasure should be, then 40 paces north means $y_0 = 40$, and 30 paces west means $x_0 = -30$:

$$(x_0,y_0) = (-30,40)$$

Because y_0 is positive, we use this formula to determine the bearing or heading α_0:

$$\begin{aligned}\alpha_0 &= \arctan\,(x_0/y_0)\\ &= \arctan\,(-30/40)\\ &= \arctan\; -0.75\\ &= -37^\circ\end{aligned}$$

This is a negative angle, so to get it into the standard form, we must add 360°:

$$\begin{aligned}\alpha_0 &= -37^\circ + 360^\circ = 360^\circ - 37^\circ\\ &= 323^\circ\end{aligned}$$

To find the value of the range, r_0, use this formula:

$$\begin{aligned}r_0 &= (x_0^2 + y_0^2)^{1/2}\\ &= (30^2 + 40^2)^{1/2}\\ &= (900 + 1600)^{1/2}\\ &= 2500^{1/2}\\ &= 50\end{aligned}$$

This means $(\alpha_0,r_0) = (323^\circ,50)$. Proceed 50 paces, approximately north by northwest. Then, if you have a shovel, go ahead and dig!

Alternative 3D Coordinates

Here are some coordinate systems that are used in mathematics and science when working in 3D space.

LATITUDE AND LONGITUDE

Latitude and *longitude* angles uniquely define the positions of points on the surface of a sphere or in the sky. The scheme for geographic locations on the earth is illustrated in Fig. 10-15A. The *polar axis* connects two specified

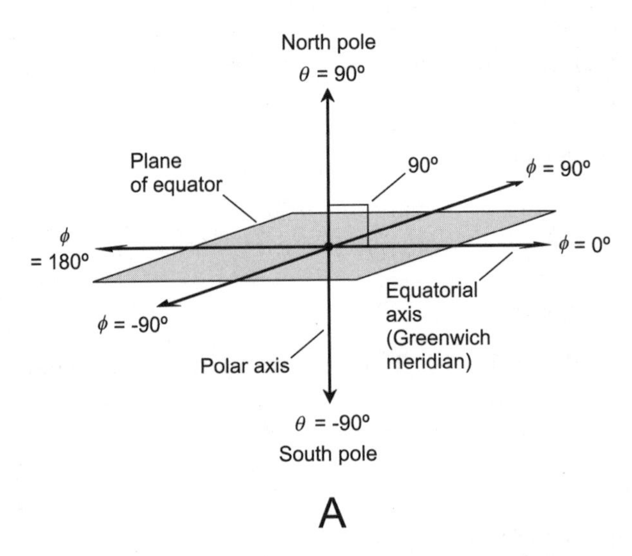

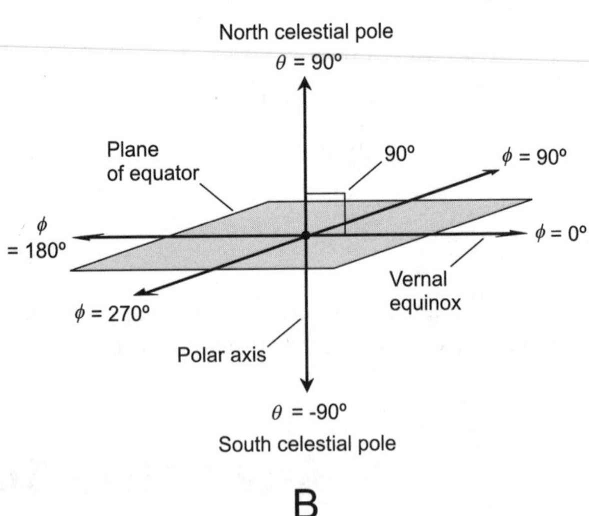

Fig. 10-15. (A) Latitude and longitude coordinates for locating points on the earth's surface. (B) Declination and right ascension coordinates for locating points in the sky.

points at antipodes on the sphere. These points are assigned latitude $\theta = 90°$ (north pole) and $\theta = -90°$ (south pole). The *equatorial axis* runs outward from the center of the sphere at a 90° angle to the polar axis. It is assigned longitude $\phi = 0°$.

Latitude θ is measured positively (north) and negatively (south) relative to the plane of the equator. Longitude ϕ is measured counterclockwise (positively) and clockwise (negatively) relative to the equatorial axis. The angles are restricted as follows:

$$-90° \leq \theta \leq 90°$$
$$-180° < \phi \leq 180°$$

On the earth's surface, the half-circle connecting the 0° longitude line with the poles passes through Greenwich, England (not Greenwich Village in New York City!) and is known as the *Greenwich meridian* or the *prime meridian.* Longitude angles are defined with respect to this meridian.

CELESTIAL COORDINATES

Celestial latitude and *celestial longitude* are extensions of the earth's latitude and longitude into the heavens. The same set of coordinates used for geographic latitude and longitude applies to this system. An object whose celestial latitude and longitude coordinates are (θ,ϕ) appears at the zenith in the sky (directly overhead) from the point on the earth's surface whose latitude and longitude coordinates are (θ,ϕ).

Declination and *right ascension* define the positions of objects in the sky relative to the stars. Figure 10-15B applies to this system. Declination (θ) is identical to celestial latitude. Right ascension (ϕ) is measured eastward from the *vernal equinox* (the position of the sun in the heavens at the moment spring begins in the northern hemisphere). The angles are restricted as follows:

$$-90° \leq \theta \leq 90°$$
$$0° \leq \phi < 360°$$

HOURS, MINUTES, AND SECONDS

Astronomers use a peculiar scheme for right ascension. Instead of expressing the angles of right ascension in degrees or radians, they use *hours, minutes,* and *seconds* based on 24 hours in a complete circle (corresponding to the 24 hours in a day). That means each hour of right ascension is equivalent to 15°. As if that isn't confusing enough, the minutes and seconds of right ascension are not the same as the fractional degree units by the same names more often encountered. One minute of right ascension is 1/60 of an hour or $^1/_4$ of a degree, and one second of right ascension is 1/60 of a minute or 1/240 of a degree.

CYLINDRICAL COORDINATES

Figures 10-16A and 10-16B show two systems of *cylindrical coordinates* for specifying the positions of points in three-space.

In the system shown in Fig. 10-16A, we start with Cartesian xyz-space. Then an angle θ is defined in the xy-plane, measured in degrees or radians (but usually radians) counterclockwise from the positive x axis, which is called the *reference axis*. Given a point P in space, consider its projection P' onto the xy-plane. The position of P is defined by the ordered triple (θ,r,h). In this ordered triple, θ represents the angle measured counterclockwise between P' and the positive x axis in the xy-plane, r represents the distance or radius from P' to the origin, and h represents the distance, called the altitude or height, of P above the xy-plane. (If h is negative, then P is below the xy-plane.) This scheme for cylindrical coordinates is preferred by mathematicians, and also by some engineers and scientists.

In the system shown in Fig. 10-16B, we again start with Cartesian xyz-space. The xy-plane corresponds to the surface of the earth in the vicinity of the origin, and the z axis runs straight up (positive z values) and down (negative z values). The angle θ is defined in the xy-plane in degrees (but never radians) *clockwise* from the positive y axis, which corresponds to geographic north. Given a point P in space, consider its projection P' onto the xy-plane. The position of P is defined by the ordered triple (θ,r,h), where θ represents the angle measured clockwise between P' and geographic north, r represents the distance or radius from P' to the origin, and h represents the altitude or height of P above the xy-plane. (If h is negative, then P is below the xy-plane.) This scheme is preferred by navigators and aviators.

SPHERICAL COORDINATES

Figures 10-17A to 10-17C show three systems of *spherical coordinates* for defining points in space. The first two are used by astronomers and aerospace scientists, while the third one is of use to navigators and surveyors.

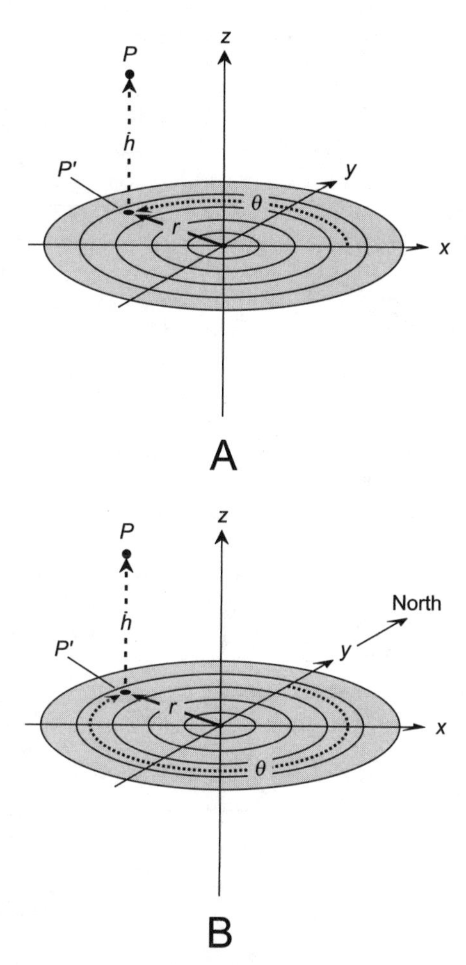

Fig. 10-16. (A) Mathematician's form of cylindrical coordinates for defining points in three-space. (B) Astronomer's and navigator's form of cylindrical coordinates for defining points in three-space.

In the scheme shown in Fig. 10-17A, the location of a point P is defined by the ordered triple (θ,ϕ,r) such that θ represents the declination of P, ϕ represents the right ascension of P, and r represents the distance or radius from P to the origin. In this example, angles are specified in degrees (except in the case of the astronomer's version of right ascension, which is expressed in hours, minutes, and seconds as defined earlier in this chapter). Alternatively, the angles can be expressed in radians. This system is fixed relative to the stars.

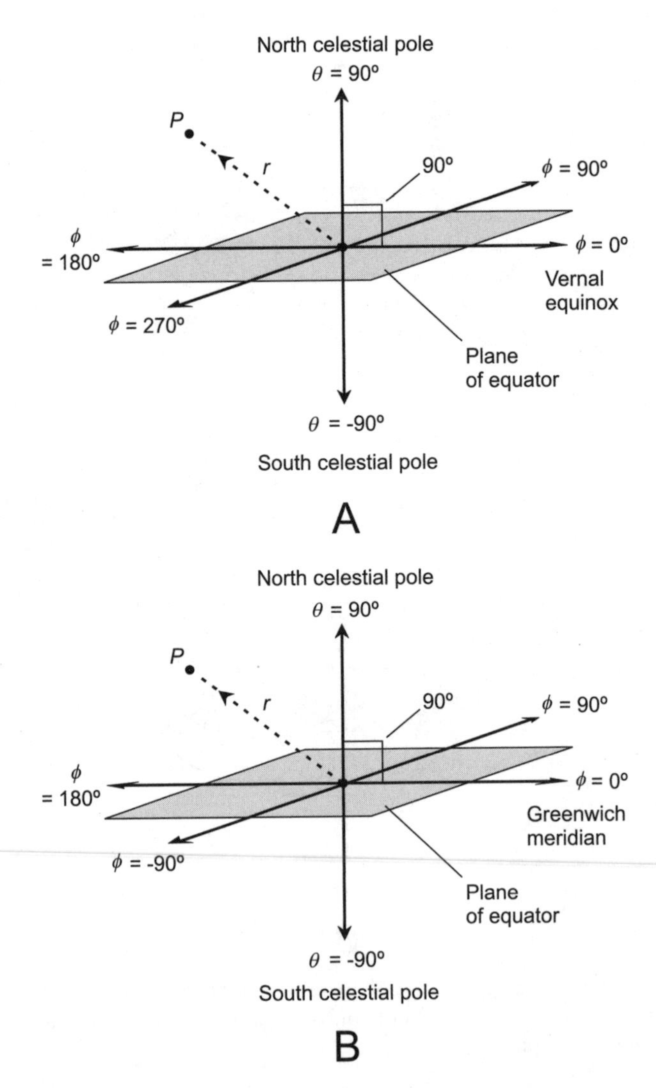

Fig. 10-17. (A) Spherical coordinates for defining points in three-space, where the angles represent declination and right ascension. (B) Spherical coordinates for defining points in three-space, where the angles represent celestial latitude and longitude.

Instead of declination and right ascension, the variables θ and ϕ can represent celestial latitude and celestial longitude respectively, as shown in Fig. 10-17B. This system is fixed relative to the earth, rather than relative to the stars.

There's yet another alternative: θ can represent elevation (the angle above the horizon) and ϕ can represent the azimuth (bearing or heading), measured

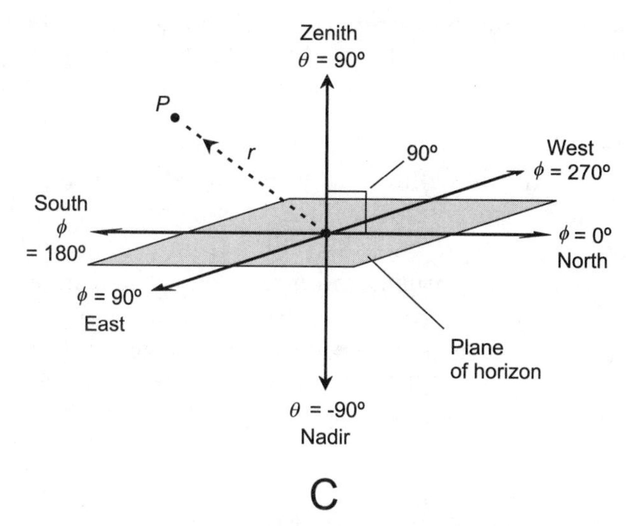

Fig. 10-17. (C) Spherical coordinates for defining points in three-space, where the angles represent elevation (angle above the horizon) and azimuth (also called bearing or heading).

clockwise from geographic north. In this case, the reference plane corresponds to the horizon, not the equator, and the elevation can range between, and including, $-90°$ (the nadir, or the point directly underfoot) and $+90°$ (the zenith, or the point directly overhead). This is shown in Fig. 10-17C. In a variant of this system, the angle θ is measured with respect to the zenith, rather than the horizon. Then the range for this angle is $0° \leq \theta \leq 180°$.

PROBLEM 10-7

What are the celestial latitude and longitude of the sun on the first day of spring, when the sun lies in the plane of the earth's equator?

SOLUTION 10-7

The celestial latitude of the sun on the first day of spring (March 21, the vernal equinox) is 0°, which is the same as the latitude of the earth's equator. The celestial longitude depends on the time of day. It is 0° (the Greenwich meridian) at high noon in Greenwich, England or any other location at 0° longitude. From there, the celestial longitude of the sun proceeds west at the rate of 15° per hour (360° per 24 hours).

PROBLEM 10-8

Suppose you stand in a huge, perfectly flat field and fly a kite on a string 500 meters long. The wind blows directly from the east. The point on the ground directly below the kite is r meters away from you, and the kite is 400 meters above the ground. If your body represents the origin and the units of a coordinate system are one meter in size, what is the position of the kite in the cylindrical coordinate scheme preferred by navigators and aviators?

SOLUTION 10-8

The position of the kite is defined by the ordered triple (θ,r,h), where θ represents the angle measured clockwise from geographic north to a point directly under the kite, r represents the radius from a point on the ground directly under the kite to the origin, and h represents the distance (height or altitude) of the kite above the ground. Because the wind blows from the east, you know that the kite must be directly west of the origin (represented by your body), so $\theta = 270°$. The kite is 400 meters off the ground, so $h = 400$. The value of r can be found by the Pythagorean theorem:

$$r^2 + 400^2 = 500^2$$

$$r^2 + 160{,}000 = 250{,}000$$

$$r^2 = 250{,}000 - 160{,}000$$

$$r^2 = 90{,}000$$

$$r = (90{,}000)^{1/2} = 300$$

Therefore, $(\theta,r,h) = (270°,300,400)$ in the system of cylindrical coordinates preferred by navigators and aviators.

Quiz

Refer to the text in this chapter if necessary. A good score is eight correct. Answers are in the back of the book.

1. In spherical coordinates, the position of a point is specified by
   (a) two angles and a distance
   (b) two distances and an angle
   (c) three distances
   (d) three angles

2. Suppose a point has the coordinates $(\theta,r) = (\pi,3)$ in the conventional (or mathematician's) polar scheme. It is implied from this that the angle is
   (a) negative
   (b) expressed in radians

(c) greater than 360°
(d) ambiguous

3. Suppose a point has the coordinates $(\theta,r) = (\pi/4,6)$ in the mathematician's polar scheme. What are the coordinates (α,r) of the point in the navigator's polar scheme?
(a) They cannot be determined without more information
(b) (–45°,6)
(c) (45°,6)
(d) (135°,6)

4. Suppose we are given the simple relation $g(x) = x$. In Cartesian coordinates, this has the graph $y = x$. What is the equation that represents the graph of this relation in the mathematician's polar coordinate system?
(a) $r = \theta$
(b) $r = 1/\theta$, where $\theta \neq 0°$
(c) $\theta = 45°$, where r can range over the entire set of real numbers
(d) $\theta = 45°$, where r can range over the set of non-negative real numbers

5. Suppose we set off on a bearing of 135° in the navigator's polar coordinate system. We stay on a straight course. If the starting point is considered the origin, what is the graph of our path in Cartesian coordinates?
(a) $y = x$, where $x \geq 0$
(b) $y = 0$, where $x \geq 0$
(c) $x = 0$, where $y \geq 0$
(d) $y = -x$, where $x \geq 0$

6. The direction angle in the navigator's polar coordinate system is measured
(a) in a clockwise sense
(b) in a counterclockwise sense
(c) in either sense
(d) only in radians

7. The graph of $r = -3\theta$ in the mathematician's polar coordinate system looks like
(a) a circle
(b) a cardioid
(c) a spiral
(d) nothing; it is undefined

8. Suppose you see a balloon hovering in the sky over a calm ocean. You are told that it is at azimuth 30°, that it is 3500 meters above the ocean surface, and that the point directly underneath it is 5000 meters away from you. This information is an example of the position of the balloon expressed in a form of
   (a) Cartesian coordinates
   (b) cylindrical coordinates
   (c) spherical coordinates
   (d) celestial coordinates

9. Suppose we are given a point and told that its Cartesian coordinate is $(x,y) = (0,-5)$. In the mathematician's polar scheme, the coordinates of this point are
   (a) $(\theta,r) = (3\pi/2,5)$
   (b) $(\theta,r) = (3\pi/2,-5)$
   (c) $(\theta,r) = (-5,3\pi/2)$
   (d) ambiguous; we need more information to specify them

10. Suppose a radar unit shows a target that is 10 kilometers away in a southwesterly direction. It is moving directly away from us. When its distance has doubled to 20 kilometers, what has happened to the x and y coordinates of the target in Cartesian coordinates? Assume we are located at the origin.
    (a) They have both increased by a factor equal to the square root of 2
    (b) They have both doubled
    (c) They have both quadrupled
    (d) We need to specify the size of each unit in the Cartesian coordinate system in order to answer this question

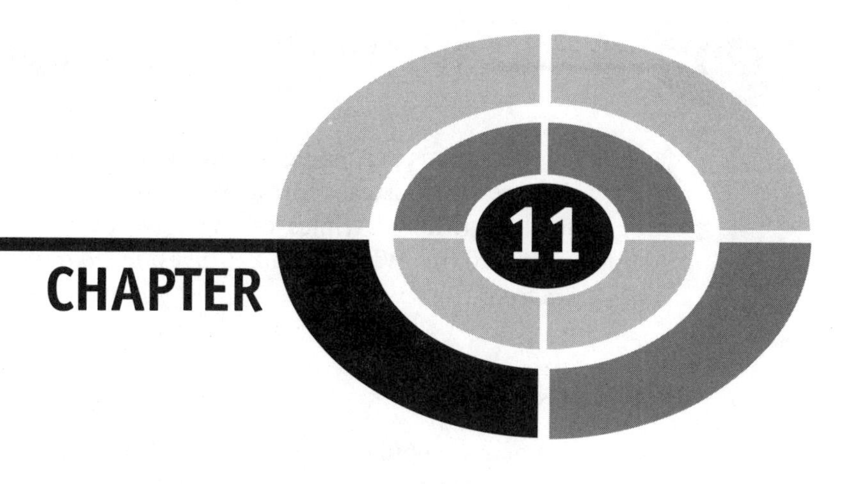

Hyperspace and Warped Space

Some of the concepts in this chapter are among the most esoteric in all of mathematics, with far-reaching applications. *Hyperspace* (space of more than three dimensions) and *warped space* can be envisioned even by young children. Some people who think they are not mathematically inclined find *non-Euclidean geometry* interesting, perhaps because some of it resembles science fiction.

Cartesian *n*-Space

As we have seen, the rectangular (or Cartesian) coordinate plane is defined by two number lines that intersect perpendicularly at their zero points. The lines form axes, often called the x axis and the y axis. Points in such a system are identified by ordered pairs of the form (x,y). The point defined by $(0,0)$ is called the origin. Cartesian three-space is defined by three number lines that intersect at a single point, corresponding to the zero point of each line, and such that each line is perpendicular to the plane determined by the other two lines. The lines form axes, representing variables such as x, y, and z. Points are defined by ordered triples of the form (x,y,z). The origin is the point defined by (0,0,0). What about *Cartesian four-space*? Or *five-space*? Or *infinity-space*?

IMAGINE THAT!

A system of rectangular coordinates in four dimensions—Cartesian four-space or 4D space—is defined by four number lines that intersect at a single point, corresponding to the zero point of each line, and such that each of the lines is perpendicular to the other three. The lines form axes, representing variables such as w, x, y, and z. Alternatively, the axes can be labeled x_1, x_2, x_3, and x_4. Points are identified by *ordered quadruples* of the form (w,x,y,z) or (x_1,x_2,x_3,x_4). The origin is defined by (0,0,0,0). As with the variables or numbers in ordered pairs and triples, there are no spaces after the commas. Everything is all scrunched together.

At first you might think, "Cartesian four-space isn't difficult to imagine," and draw an illustration such as Fig. 11-1 to illustrate it. But when you start trying to plot points in this system, you'll find out there is a problem. You can't define points in such a rendition of four-space without ambiguity. There

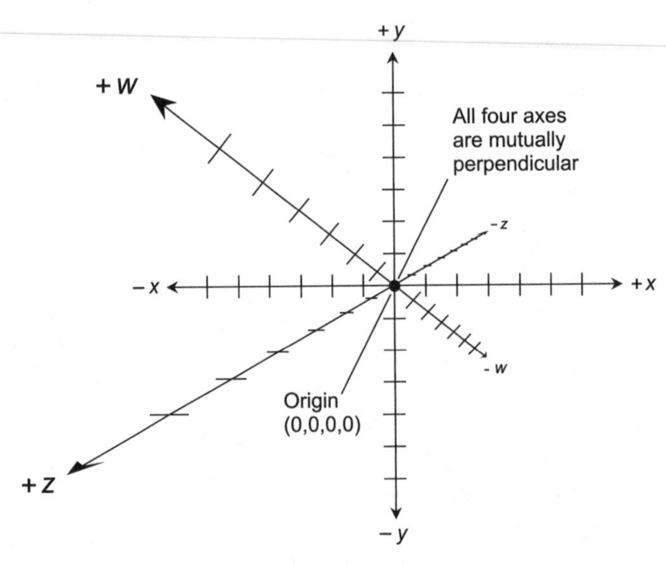

Fig. 11-1. Concept of Cartesian four-space. The w, x, y, and z axes are all mutually perpendicular at the origin point (0,0,0,0).

are too many possible values of the ordered quadruple (w,x,y,z), and not enough points in 3D space to accommodate them all. In 3D space as we know it, four number lines such as those shown in Fig. 11-1 cannot be oriented so they intersect at a single point with all four lines perpendicular to the other three.

Imagine the point in a room where the walls meet the floor. Unless the building has an unusual architecture or is sagging because of earth movement, there are three straight line segments defined by this intersection. One of the line segments runs up and down between the two walls, and the other two run horizontally between the two walls and the floor. The line segments are all mutually perpendicular at the point where they intersect. They are like the x, y, and z axes in Cartesian three-space. Now think of a fourth line segment that has one end at the intersection point of the existing three line segments, and that is perpendicular to them all. Such a line segment can't exist in ordinary space! But in four dimensions, or hyperspace, it can. If you were a 4D creature, you would not be able to understand how 3D creatures could possibly have trouble envisioning four line segments all coming together at mutual right angles.

Mathematically, we can work with Cartesian four-space, even though it cannot be directly visualized. This makes 4D geometry a powerful mathematical tool. As it turns out, the universe we live in requires four or more dimensions in order to be fully described. Albert Einstein was one of the first scientists to put forth the idea that the "fourth dimension" exists.

TIME-SPACE

You've seen *time lines* in history books. You've seen them in graphs of various quantities, such as temperature, barometric pressure, or the Dow Jones industrial average plotted as functions of time. Isaac Newton, one of the most renowned mathematicians in the history of the Western world, imagined time as flowing smoothly and unalterably. Time, according to Newtonian physics, does not depend on space, nor space on time.

Wherever you are, however fast or slow you travel, and no matter what else you do, the cosmic clock, according to Newtonian (or classical) physics, keeps ticking at the same absolute rate. In most practical scenarios, this model works quite well; its imperfections are not evident. It makes the time line a perfect candidate for a "fourth perpendicular axis." Nowadays we know that Newton's model represents an oversimplification; some folks might say it is conceptually flawed. But it is a good approximation of reality under most everyday circumstances.

Mathematically, we can envision a time line passing through 3D space, perpendicular to all three spatial axes such as the intersections between two walls and the floor of a room. The time axis passes through three-space at some chosen origin point, such as the point where two walls meet the floor in a room, or the center of the earth, or the center of the sun, or the center of the Milky Way galaxy.

In four-dimensional (4D) *Cartesian time-space* (or simply *time-space*), each point follows its own time line. Assuming none of the points is in motion with respect to the origin, all the points follow time lines parallel to all the other time lines, and they are all constantly perpendicular to three-space. Dimensionally reduced, this sort of situation can be portrayed as shown in Fig. 11-2.

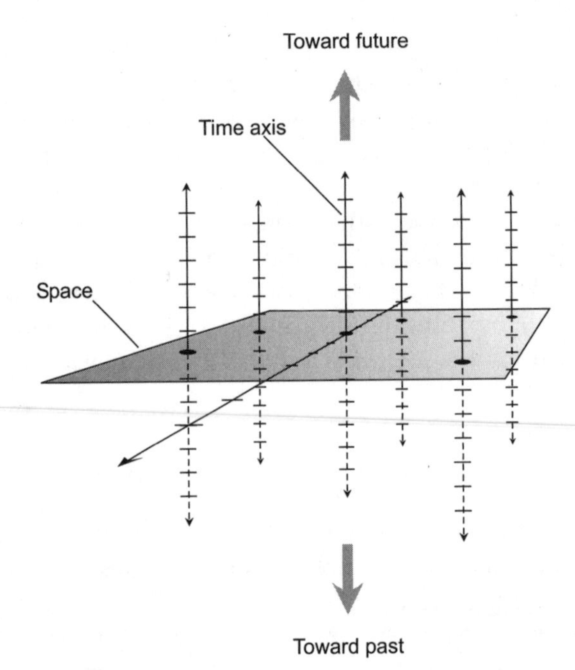

Fig. 11-2. Time as a fourth dimension. Three-space is shown dimensionally reduced, as a plane. Each stationary point in space follows a time line perpendicular to 3D space and parallel to the time axis.

POSITION VS MOTION

Things get more interesting when we consider the paths of moving points in time-space. Suppose, for example, that we choose the center of the sun as the origin point for a Cartesian three-space coordinate system.

Imagine that the x and y axes lie in the plane of the earth's orbit around the sun. Suppose the positive x axis runs from the sun through the earth's position in space on March 21, and thence onward into deep space (roughly towards the constellation Virgo for you astronomy buffs). Then the negative x axis runs from the sun through the earth's position on September 21 (roughly through Pisces), the positive y axis runs from the sun through the earth's position on June 21 (roughly toward the constellation Sagittarius), and the negative y axis runs from the sun through the earth's position on December 21 (roughly toward Gemini). The positive z axis runs from the sun toward the north celestial pole (in the direction of Polaris, the North Star), and the negative z axis runs from the sun toward the south celestial pole. Let each division on the coordinate axes represent one-quarter of an *astronomical unit* (AU), where 1 AU is defined as the mean distance of the earth from the sun (about 150,000,000 kilometers). Figure 11-3A shows this coordinate system, with the earth on the positive x axis, at a distance of 1 AU. The coordinates of the earth at this time are (1,0,0) in the xyz-space we have defined.

Of course, the earth doesn't remain fixed. It orbits the sun. Let's take away the z axis in Fig. 11-3A and replace it with a time axis called t. What will the earth's path look like in xyt-space, if we let each increment on the t axis represent exactly one-quarter of a year? The earth's path through this dimensionally-reduced time-space is not a straight line, but instead is a helix as shown in Fig. 11-3B. The earth's distance from the t axis remains nearly constant, although it varies a little because the earth's orbit around the sun is not a perfect circle. Every quarter of a year, the earth advances 90° around the helix.

Some Hyper Objects

Now that we're no longer bound to 3D space, let's put our newly empowered imaginations to work. What are 4D objects like? How about five dimensions (5D) and beyond?

TIME AS DISPLACEMENT

When considering time as a dimension, it is convenient to have some universal standard that relates time to spatial displacement. How many kilometers are there in one second of time? At first this seems like a ridiculous

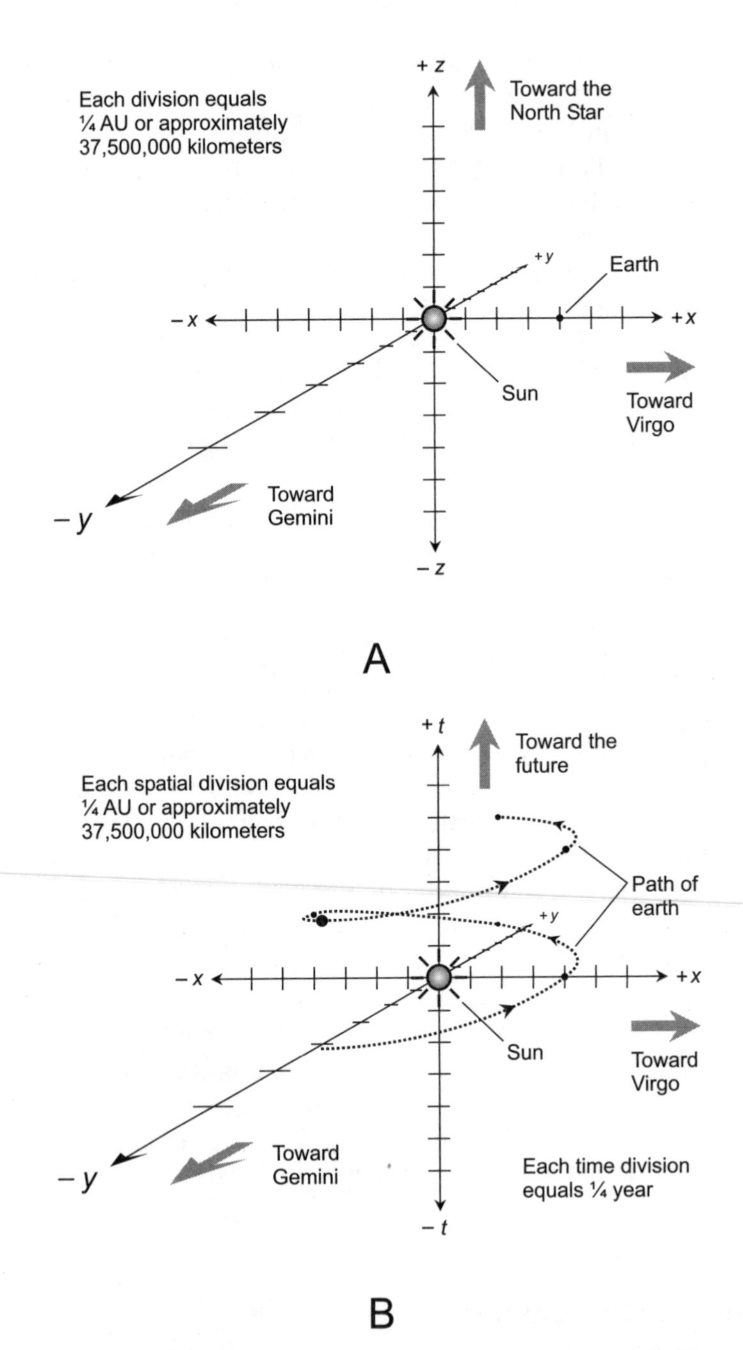

Fig. 11-3. (A) A Cartesian coordinate system for the position of the earth in 3D space. (B) A dimensionally reduced Cartesian system for rendering the path of the earth through 4D time-space.

question, akin to asking how many apples there are in a gallon of gasoline. But think of it like this: time and displacement can be related by speed in a sensible way, as long as the speed is known and is constant.

Suppose someone tells you, "Jimsville is an hour away from Joesville." You've heard people talk like this, and you understand what they mean. A certain speed is assumed. How fast must you go to get from Jimsville to Joesville in an hour? If Jimsville and Joesville are 50 kilometers from each other, then you must travel 50 kilometers per hour in order to say that they are an hour apart. If they are only 20 kilometers apart, then you need only travel 20 kilometers per hour to make the same claim.

Perhaps you remember the following formula from elementary physics:

$$d = st$$

where d is the distance in kilometers, s is the speed of an object in kilometers per hour, and t is the number of hours elapsed. Using this formula, it is possible to define time in terms of displacement and vice versa.

UNIVERSAL SPEED

Is there any speed that is universal, and that can be used on that basis as an absolute relating factor between time and displacement? Yes, according to Albert Einstein's famous relativity theory. The speed of light in a vacuum, commonly denoted c, is constant, and it is independent of the point of view of the observer (as long as the observer is not accelerating at an extreme rate or in a super-intense gravitational field). This constancy of the speed of light is a fundamental principle of the theory of special relativity. The value of c is very close to 299,792 kilometers per second; let's round it off to 300,000 kilometers per second. If d is the distance in kilometers and t is the time in seconds, the following formula is absolute in a certain cosmic sense:

$$d = ct = 300{,}000t$$

According to this model, the moon, which is about 400,000 kilometers from the earth, is 1.33 *second-equivalents* distant. The sun is about 8.3 *minute-equivalents* away. The Milky Way galaxy is 100,000 *year-equivalents* in diameter. (Astronomers call these units *light-seconds*, *light-minutes*, and *light-years*.) We can also say that any two points in time that are separated by one second, but that occupy the same xyz coordinates in Cartesian three-space, are separated by 300,000 *kilometer-equivalents* along the t axis.

At this instant yesterday, if you were in the same location as you are now, your location in time-space was 24 (hours per day) × 60 (minutes per hour) ×

60 (seconds per minute) × 300,000 (kilometers per second), or 25,920,000,000 kilometer-equivalents away. This mode of thinking takes a bit of getting used to. But after a while, it starts to make sense, even if it's a slightly perverse sort of sense. It is, for example, just about as difficult to jump 25,920,000,000 kilometers in a single bound, as it is to change what happened in your room at this time yesterday.

The above formula can be modified for smaller distances. If d is the distance in kilometers and t is the time in milliseconds (units of 0.001 of a second), then:

$$d = 300t$$

This formula also holds for d in meters and t in microseconds (units of 0.000001, or 10^{-6}, second), and for d in millimeters (units of 0.001 meter) and t in nanoseconds (units of 0.000000001, or 10^{-9}, second). Thus we might speak of *meter-equivalents*, *millimeter-equivalents*, *microsecond-equivalents*, or *nanosecond-equivalents*.

THE FOUR-CUBE

Imagine some of the simple, regular polyhedra in Cartesian four-space. What are their properties? Think about a *four-cube*, also known as a *tesseract*. This is an object with several identical 3D *hyperfaces*, all of which are cubes. How many vertices does a tesseract have? How many edges? How many 2D faces? How many 3D hyperfaces? How can we envision such a thing to figure out the answers to these questions?

This is a situation in which time becomes useful as a fourth spatial dimension. We can't make a 4D model of a tesseract out of toothpicks to examine its properties, and few people (if any) can envision such a thing. But we can imagine a cube that pops into existence for a certain length of time and then disappears a little later, such that it "lives" for a length of time equivalent to the length of any of its spatial edges, and does not move during its existence. Because we have defined an absolute relation between time and displacement, we can graph a tesseract in which each edge is, say, 300,000 kilometer-equivalents long. It is an ordinary 3D cube that measures 300,000 kilometers along each edge. It pops into existence at a certain time t_0 and then disappears 1 second later, at $t_0 + 1$. The sides of the cube are each 1 second-equivalent in length, and the cube "lives" for 300,000 kilometer-equivalents of time.

Figure 11-4A shows a tesseract in dimensionally reduced form. Each division along the x and y axes represents 100,000 kilometers (the equivalent of 1/3 second), and each division along the t axis represents 1/3 second (the

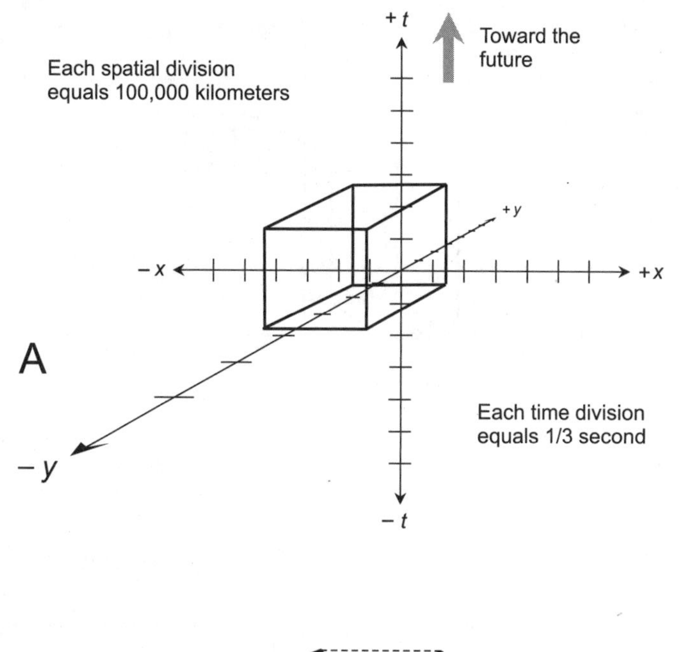

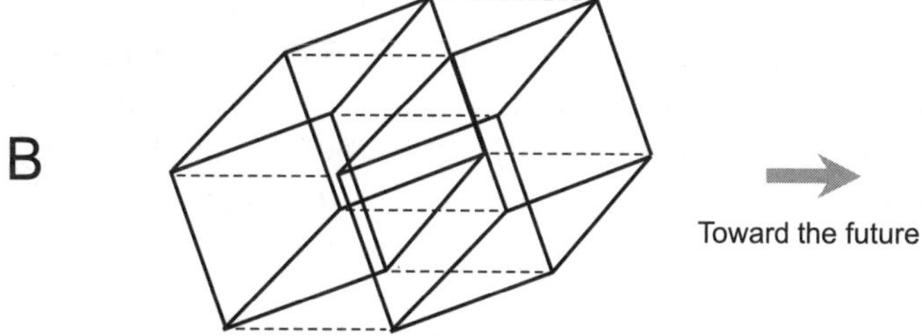

Fig. 11-4. At A, a dimensionally reduced plot of a time-space tesseract. At B, another rendition of a tesseract, portraying time as lateral motion.

equivalent of 100,000 kilometers). Figure 11-4B is another rendition of this object, illustrated as two 3D cubes (in perspective) connected by dashed lines representing the passage of time.

THE RECTANGULAR FOUR-PRISM

A tesseract is a special form of the more general figure, known as a *rectangular four-prism* or *rectangular hyperprism*. Such an object is a 3D rectangular prism that abruptly comes into existence, lasts a certain length of time, disappears all at once, and does not move during its "lifetime." Figure 11-5

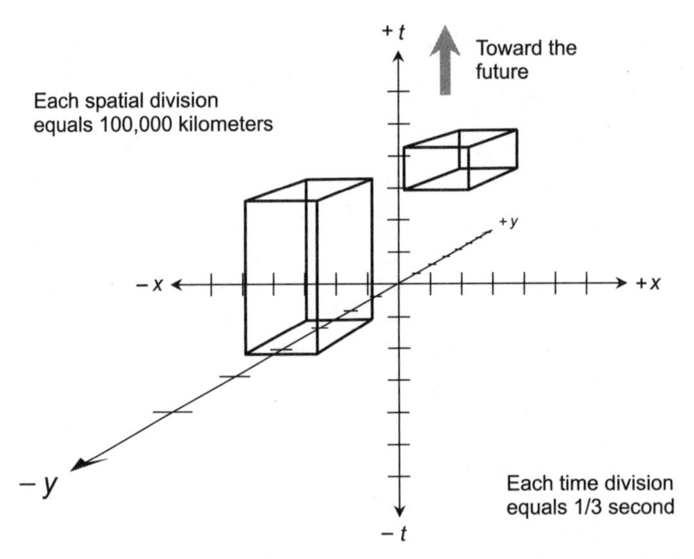

Fig. 11-5. Dimensionally reduced plots of two rectangular hyperprisms in time-space.

shows two examples of rectangular four-prisms in dimensionally reduced time-space.

Suppose the height, width, depth, and lifetime of a rectangular hyperprism, all measured in kilometer-equivalents, are h, w, d, and t, respectively. Then the *4D hypervolume* of this object (call it V_{4D}), in *quartic kilometer-equivalents*, is given by the product of them all:

$$V_{4D} = hwdt$$

The mathematics is the same if we express the height, width, depth, and lifetime of the object in second-equivalents; the 4D hypervolume is then equal to the product $hwdt$ in *quartic second-equivalents.*

IMPOSSIBLE PATHS

Certain paths are impossible in Cartesian 4D time-space as we've defined it here. According to Einstein's special theory of relativity, nothing can travel faster than the speed of light. This restricts the directions in which line segments, lines, and rays can run when denoting objects in motion.

Consider what happens in 4D Cartesian time-space when a light bulb is switched on. Suppose the bulb is located at the origin, and is surrounded by millions of kilometers of empty space. When the switch is closed and the bulb is first illuminated, *photons* (particles of light) emerge. These initial, or leading, photons travel outward from the bulb in expanding spherical paths. If we dimensionally reduce this situation and graph it, we get an expanding circle centered on the time axis, which, as time passes, generates a cone as shown in Fig. 11-6. In true 4D space this is a *hypercone* or *four-cone*. The surface of the four-cone is 3D: two spatial dimensions and one time dimension.

Imagine an object that starts out at the location of the light bulb, and then moves away from the bulb as soon as the bulb is switched on. This object must follow a path entirely within the *light cone* defined by the initial photons from the bulb. Figure 11-6 shows one plausible path and one implausible path.

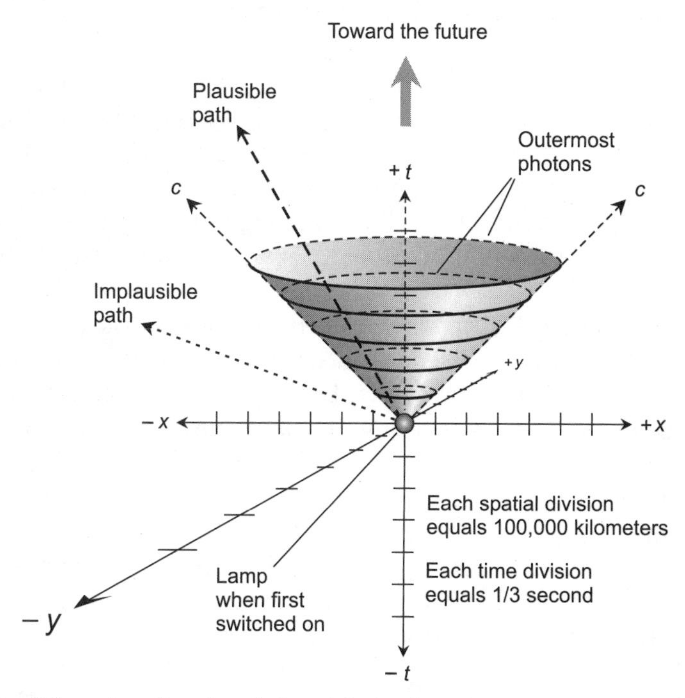

Fig. 11-6. Dimensionally reduced plot of the leading photons from a light bulb. Paths outside the light cone represent speeds greater than c (the speed of light) and are therefore implausible.

GENERAL TIME-SPACE HYPERVOLUME

Suppose there is an object—any object—in 3D space. Let its spatial volume in cubic kilometer-equivalents be equal to V_{3D}. Suppose that such an object pops into existence, lasts a certain length of time t in kilometer-equivalents, and then disappears. Suppose that this object does not move with respect to you, the observer, during its "lifetime." Then its 4D time-space hypervolume V_{4D} is given by this formula:

$$V_{4D} = V_{3D}t$$

That is to say, the 4D time-space hypervolume of any object is equal to its spatial volume multiplied by its lifetime, provided the time and displacement are expressed in equivalent units, and as long as there is no motion involved.

If an object moves, then a correction factor must be included in the above formula. This correction factor does not affect things very much as long as the speed of the object, call it s, is small compared with the speed of light c. But if s is considerable, the above formula becomes:

$$V_{4D} = V_{3D}t\,(1 - s^2/c^2)^{1/2}$$

The correction factor, $(1 - s^2/c^2)^{1/2}$, is close to 1 when s is a small fraction of c, and approaches 0 as s approaches c. This correction factor derives from the special theory of relativity. (It can be proven with the help of the Pythagorean theorem. The proof is not complicated, but getting into it here would take us off the subject. Let's just say that objects are "spatially squashed" at extreme speeds, and leave it at that.)

In this context, speed s is always relative. It depends on the point of view from which it is observed, witnessed, or measured. For speed to have meaning, we must always add the qualifying phrase "relative to a certain observer." In these examples, we envision motion as taking place relative to the origin of a 3D Cartesian system, which translates into lines, line segments, or rays pitched at various angles with respect to the time axis in a 4D time-space Cartesian system.

If you're still confused about kilometer-equivalents and second-equivalents, you can refer to Table 11-1 for reference. Keep in mind that time and displacement are related according to the speed of light:

$$d = ct$$

where d is the displacement (in linear units), t is the time (in time units), and c is the speed of light in linear units per unit time. Using this conversion formula, you can convert any displacement unit to an equivalent time interval, and any time unit to an equivalent displacement.

PROBLEM 11-1

How many second-equivalents are there in 1 kilometer?

SOLUTION 11-1

We know that the speed of light is 300,000 kilometers per second, so it takes 1/300,000 of a second for light to travel 1 kilometer. That is approximately

Table 11-1. Some displacement and time equivalents. Displacement equivalents are accurate to three significant figures.

DISPLACEMENT EQUIVALENT	TIME EQUIVALENT
9,460,000,000,000 kilometers	1 year
25,900,000,000 kilometers	1 solar day
1,079,000,000 kilometers	1 hour
150,000,000 kilometers (1 astronomical unit)	500 seconds (8 minutes, 20 seconds)
18,000,000 kilometers	1 minute
300,000 kilometers	1 second
300 kilometers	0.001 second (1 millisecond)
1 kilometer	0.00000333 second (3.33 microseconds)
300 meters	0.000001 second (1 microsecond)
1 meter	0.00000000333 second (3.33 nanoseconds)
300 millimeters	0.000000001 second (1 nanosecond)
1 millimeter	0.00000000000333 second (3.33 picoseconds)

0.00000333 seconds or 3.33 microseconds. One kilometer is 0.00000333 second-equivalents, or 3.33 microsecond-equivalents.

Beyond Four Dimensions

There is no limit to the number of dimensions that can be defined using the Cartesian scheme. There can be any positive whole number of dimensions. Time can be (but does not have to be) one of them.

CARTESIAN *N*-SPACE

A system of rectangular coordinates in five dimensions defines *Cartesian five-space*. There are five number lines, all of which intersect at a point corresponding to the zero point of each line, and such that each of the lines is perpendicular to the other four. The resulting axes can be called v, w, x, y, and z. Alternatively they can be called x_1, x_2, x_3, x_4, and x_5. Points are identified by *ordered quintuples* such as (v,w,x,y,z) or (x_1,x_2,x_3,x_4,x_5). The origin is defined by (0,0,0,0,0).

A system of rectangular coordinates in *Cartesian n-space* (where n is any positive integer) consists of n number lines, all of which intersect at their zero points, such that each of the lines is perpendicular to all the others. The axes can be named x_1,x_2,x_3, ..., and so on up to x_n. Points in Cartesian n-space can be uniquely defined by ordered n-tuples of the form $(x_1,x_2,x_3,\ldots,x_n)$.

Imagine a tesseract or a rectangular four-prism that pops into existence at a certain time, does not move, and then disappears some time later. This object is a *rectangular five-prism*. If x_1, x_2, x_3, and x_4 represent four spatial dimensions (in kilometer-equivalents or second-equivalents) of a rectangular four-prism in Cartesian four-space, and if t represents its "lifetime" in the same units, then the *5D hypervolume* (call it V_{5D}) is equal to the product of them all:

$$V_{5D} = x_1 x_2 x_3 x_4 t$$

This holds only as long as there is no motion. If there is motion, then the relativistic correction factor must be included.

DIMENSIONAL CHAOS

There is nothing to stop us from dreaming up a *Cartesian 25-space* in which the coordinates of the points are ordered 25-tuples $(x_1,x_2,x_3,\ldots,x_{25})$, none of which are time. Alternatively, such a hyperspace might have 24 spatial dimensions and one time dimension. Then the coordinates of a point would be defined by the ordered 25-tuple $(x_1,x_2,x_3,\ldots,x_{24},t)$.

Some cosmologists—scientists who explore the origin, structure, and evolution of the cosmos—have suggested that our universe was "born" with 11 dimensions. According to this hypothesis, not all of these dimensions can be represented by Cartesian coordinates. Some of the axes are "curled up" or *compactified* as if wrapped around tiny bubbles. Some mathematicians have played with objects that seem to be 2D in some ways and 3D in other ways. How many dimensions are there in the complicated surface of a theoretical *foam*, assuming each individual bubble is a sphere of arbitrarily tiny size and with an infinitely thin 2D surface? Two dimensions? In a way. Three? In a way. How about two and a half?

The examples we have looked at here are among the simplest. Imagine the possible ways in which a *4D parallelepiped* might exist, or a *4D sphere*. How about a *5D sphere*, or a *7D ellipsoid*? Let your mind roam free. But don't think about this stuff while driving, operating heavy equipment, cycling, or walking across a street in traffic.

DISTANCE FORMULAS

In n-dimensional Cartesian space, the shortest distance between any two points can be found by means of a formula similar to the distance formulas for 2D and 3D space. The distance thus calculated represents the length of a straight line segment connecting the two points.

Suppose there are two points in Cartesian n-space, defined as follows:

$$P = (x_1, x_2, x_3, \ldots, x_n)$$
$$Q = (y_1, y_2, y_3, \ldots y_n)$$

The length of the shortest possible path between points P and Q, written $|PQ|$, is equal to either of the following:

$$|PQ| = [(y_1 - x_1)^2 + (y_2 - x_2)^2 + (y_3 - x_3)^2 + \cdots + (y_n - x_n)^2]^{1/2}$$
$$|PQ| = [(x_1 - y_1)^2 + (x_2 - y_2)^2 + (x_3 - y_3)^2 + \cdots + (x_n - y_n)^2]^{1/2}$$

PROBLEM 11-2

Find the distance $|PQ|$ between the points $P = (4,-6,-3,0)$ and $Q = (-3,5,0,8)$ in Cartesian four-space. Assume the coordinate values to be exact; express the answer to two decimal places.

SOLUTION 11-2

Assign the numbers in these ordered quadruples the following values:

$$x_1 = 4 \qquad y_1 = -3$$
$$x_2 = -6 \qquad y_2 = 5$$
$$x_3 = -3 \qquad y_3 = 0$$
$$x_4 = 0 \qquad y_4 = 8$$

Then plug these values into either of the above two distance formulas. Let's use the first formula:

$$\begin{aligned}|PQ| &= \{(-3-4)^2 + [5-(-6)]^2 + [0-(-3)]^2 + (8-0)^2\}^{1/2} \\ &= [(-7)^2 + 11^2 + 3^2 + 8^2]^{1/2} \\ &= (49 + 121 + 9 + 64)^{1/2} \\ &= 243^{1/2} \\ &= 15.59\end{aligned}$$

PROBLEM 11-3

How many vertices are there in a tesseract?

SOLUTION 11-3

Imagine a tesseract as a 3D cube that lasts for a length of time equivalent to the linear span of each edge. When we think of a tesseract this way, and if we think of time as flowing upward from the past toward the future, the tesseract has a "bottom" that represents the instant it is "born," and a "top" that represents the instant it "dies." Both the "bottom" and the "top" of the tesseract, thus defined, are cubes. We know that a cube has eight vertices. In the tesseract, there are twice this many vertices. The eight vertices of the "bottom" cube and the eight vertices of the "top" cube are connected with line segments that run through time.

Another way to envision this is to portray a tesseract as a cube-within-a-cube (Fig. 11-7). This is one of the most popular ways that illustrators try to draw this strange 4D figure. It isn't a true picture, of course, because the "inner" and the "outer" cubes in a real tesseract are the same size. But this rendition demonstrates that there are 16 vertices in the tesseract. Look at Fig. 11-7 and count them!

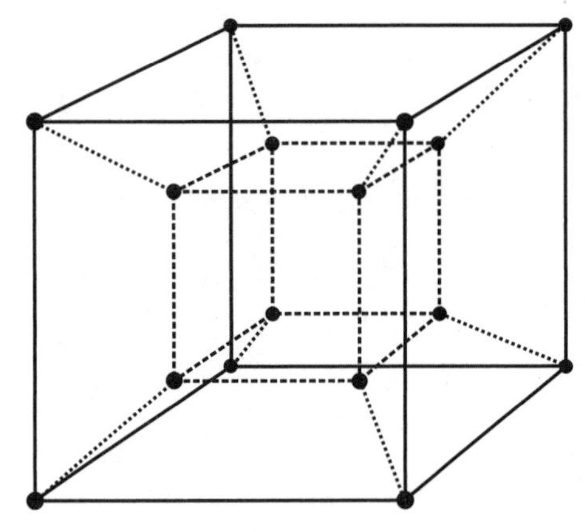

Fig. 11-7. The cube-within-a-cube portrayal of a tesseract shows that the figure has 16 vertices.

PROBLEM 11-4

What is the 4D hypervolume, V_{4D}, of a rectangular four-prism consisting of a 3D cube measuring exactly 1 meter on each edge, that "lives" for exactly 1 second, and that does not move? Express the answer in quartic kilometer-equivalents and in quartic microsecond-equivalents.

SOLUTION 11-4

We must find the 4D hypervolume of a 3D cube measuring $1 \times 1 \times 1$ meter (whose 3D volume is therefore 1 cubic meter) that "lives" for 1 second of time.

To solve the first half of this problem, note that light travels 300,000 kilometers in one second, so the four-prism "lives" for 300,000, or 10^5, kilometer-equivalents. That can be considered its length. Its cross section is a cube measuring 1 meter, or 0.001 kilometer, on each edge, so the 3D volume of this cube is $0.001 \times 0.001 \times 0.001 = 0.000000001 = 10^{-9}$ cubic kilometers. Therefore, the 4D hypervolume (V_{4D}) of the rectangular four-prism in quartic kilometer-equivalents is:

$$\begin{aligned} V_{4D} &= 300{,}000 \times 0.000000001 \\ &= 3 \times 10^5 \times 10^{-9} \\ &= 3 \times 10^{-4} \\ &= 0.0003 \text{ quartic kilometer-equivalents} \end{aligned}$$

To solve the second half of the problem, note that in 1 microsecond, light travels 300 meters, so it takes light 1/300 of a microsecond to travel 1 meter. The 3D volume of the cube in cubic microsecond-equivalents is therefore $(1/300)^3 = 1/27{,}000{,}000 = 0.00000003704 = 3.704 \times 10^{-8}$. The cube lives for 1 second, which is 1,000,000, or 10^6, microseconds. Therefore, the 4D hypervolume V_{4D} of the rectangular four-prism in quartic microsecond-equivalents is:

$$\begin{aligned} V_{4D} &= 0.00000003704 \times 1{,}000{,}000 \\ &= 3.704 \times 10^{-8} \times 10^6 \\ &= 3.704 \times 10^{-2} \\ &= 0.03704 \text{ quartic microsecond-equivalents} \end{aligned}$$

PROBLEM 11-5

Suppose the four-prism described in the previous problem moves, during its brief lifetime, at a speed of 270,000 kilometers per second relative to an observer. What is its 4D hypervolume (V_{4D}) as seen by that observer?

Express the answer in quartic kilometer-equivalents and in quartic microsecond-equivalents.

SOLUTION 11-5

The object moves at 270,000/300,000, or 9/10, of the speed of light relative to the observer. If we let s represent its speed, then $s/c = 0.9$, and $s^2/c^2 = 0.81$. We must multiply the answers to the previous problem by the following factor:

$$\begin{aligned}&(1 - s^2/c^2)^{1/2}\\&= (1 - 0.81)^{1/2}\\&= 0.19^{1/2}\\&= 0.436\end{aligned}$$

This gives us:

$$V_{4D} = 0.0003 \times 0.436 = 0.0001308 \text{ quartic kilometer-equivalents}$$

$$V_{4D} = 0.03704 \times 0.436 = 0.01615 \text{ quartic microsecond-equivalents}$$

Parallel Principle Revisited

Conventional geometry is based on five *axioms*, also called *postulates*, that were first stated by a Greek mathematician named Euclid who lived in the 3rd century B.C. Everything we have done in this book so far—even the theoretical problems involving four dimensions—has operated according to Euclid's five axioms. We have been dealing exclusively with *Euclidean geometry*. That is about to change.

EUCLID'S AXIOMS

Let's state explicitly the things Euclid believed were self-evident truths. Euclid's original wording has been changed slightly, in order to make the passages sound more contemporary. Examples of each postulate are shown in Fig. 11-8.

- Any two points P and Q can be connected by a straight line segment (Fig. 11-8A)
- Any straight line segment can be extended indefinitely and continuously to form a straight line (Fig. 11-8B)

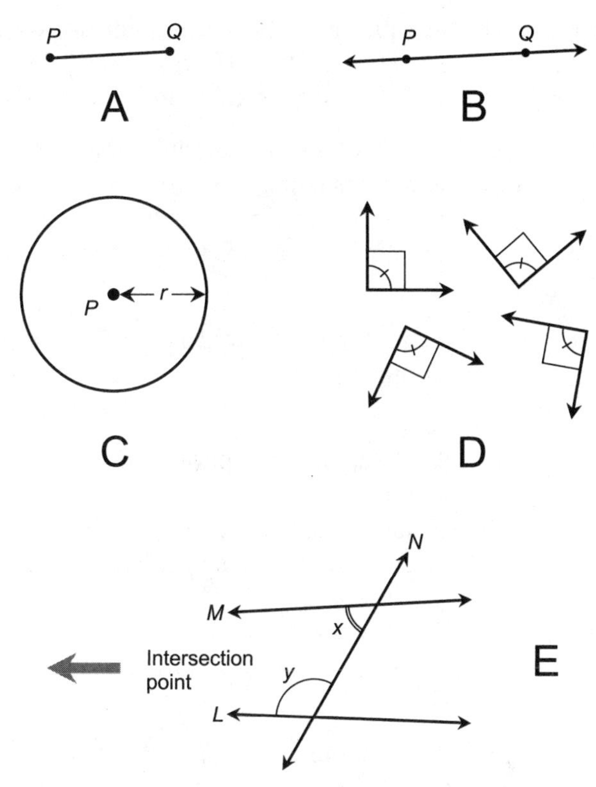

Fig. 11-8. Euclid's original five axioms. See text for discussion.

- Given any point P, a circle can be defined that has that point as its center and that has a specific radius r (Fig. 11-8C)
- All right angles are congruent; that is, they have equal measure (Fig. 11-8D)
- Suppose two lines L and M lie in the same plane and both lines are crossed by a transversal line N. Suppose the measures of the adjacent interior angles x and y sum up to less than 180° (π rad). Then lines L and M intersect on the same side of line N as angles x and y are defined (Fig. 11-8E)

THE PARALLEL POSTULATE

The last axiom stated above is known as *Euclid's fifth postulate*. It is logically equivalent to the following statement that has become known as the *parallel postulate*:

- Let L be a straight line, and let P be some point not on L. Then there exists one and only one straight line M, in the plane defined by line L and point P, that passes through point P and that is parallel to line L

This axiom—and in particular its truth or untruth—has received enormous attention. If the parallel postulate is denied, the resulting system of geometry still works. People might find it strange, but it is logically sound! Geometry doesn't need the parallel postulate. There are two ways in which the parallel postulate can be denied:

- There is no line M through point P that is parallel to line L
- There are two or more lines M_1, M_2, M_3, ... through point P that are parallel to line L

When either of these postulates replaces the parallel postulate, we are dealing with a system of non-Euclidean geometry. In the 2D case, it is a *non-Euclidean surface*. Visually, such a surface looks warped or curved. There are, as you can imagine, infinitely many ways in which a surface can be non-Euclidean.

GEODESICS

In a non-Euclidean universe, the concept of "straightness" must be modified. Instead of thinking about "straight lines" or "straight line segments," we must think about *geodesics*.

Suppose there are two points P and Q on a non-Euclidean surface. The *geodesic segment* or *geodesic arc* connecting P and Q is the set of points representing the shortest possible path between P and Q that lies on the surface. If the geodesic arc is extended indefinitely in either direction on the surface beyond P and Q, the result is a *geodesic*.

The easiest way to imagine a geodesic arc is to think about the path that a thin ray of light would travel between two points, if confined to a certain 2D universe. The extended geodesic is the path that the ray would take if allowed to travel over the surface forever without striking any obstructions. On the surface of the earth, a geodesic arc is the path that an airline pilot takes when flying from one place to another far away, such as from Moscow, Russia to Tokyo, Japan (neglecting takeoff and landing patterns, and any diversions necessary to avoid storms or hostile air space).

When we re-state the parallel postulate as it applies to both Euclidean and non-Euclidean surfaces, we must replace the term "line" with "geodesic." Here is the parallel postulate given above, modified to cover all contingencies.

MODIFIED PARALLEL POSTULATE

Two geodesics G and H on a given surface are parallel if and only if they do not intersect at any point. Let G be a geodesic, let X be a surface, and let P be some point not on geodesic G. Then one of the following three situations holds true:

- There is exactly one geodesic H on the surface X through point P that is parallel to geodesic G
- There is no geodesic H on the surface X through point P that is parallel to geodesic G
- There are two or more geodesics H_1, H_2, H_3, ... on the surface X through point P that are parallel to geodesic G

NO PARALLEL GEODESICS

Now imagine a universe in which there is no such thing as a pair of parallel geodesics. In this universe, if two geodesics that "look" parallel on a local scale are extended far enough, they eventually intersect. This type of non-Euclidean geometry is called *elliptic geometry*. It is also known as *Riemannian geometry*, named after Georg Riemann, a German mathematician who lived from 1826 until 1866 and who was one of the first mathematicians to recognize that geometry doesn't have to be Euclidean.

A universe in which there are no pairs of parallel geodesics is said to have *positive curvature*. A surface with positive curvature is warped in the same sense, no matter how the axis is oriented. Examples of 2D universes with positive curvature are the surfaces of spheres, oblate (flattened) spheres, and ellipsoids.

Figure 11-9 is an illustration of a sphere with a triangle and a quadrilateral on the surface. The sides of polygons in non-Euclidean geometry are always geodesic arcs, just as, in Euclidean geometry, they are always straight line segments. The interior angles of the triangle and the quadrilateral add up to more than 180° and 360°, respectively. The measures of the interior angles of an n-sided polygon on a Riemannian surface always sum up to more than the sum of the measures of the interior angles of an n-sided polygon on a flat plane.

On the surface of the earth, all the lines of longitude, called *meridians*, are geodesics. So is the equator. But latitude circles other than the equator, called *parallels*, are not geodesics. For example, the equator and the parallel representing 10° north latitude do not intersect, but they are not both geodesics.

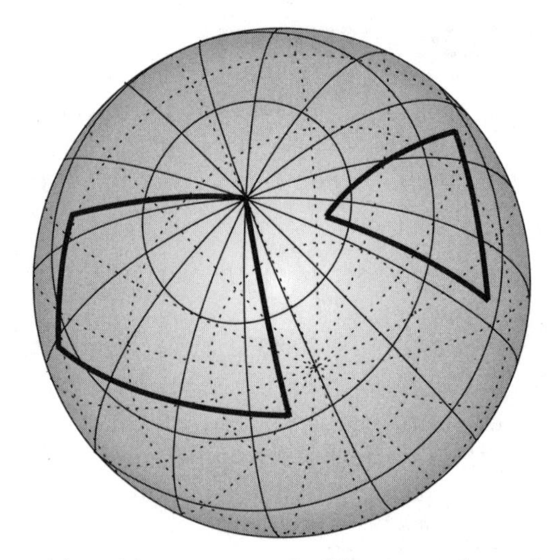

Fig. 11-9. A surface with positive curvature, in this case a sphere, showing a triangle and a quadrilateral whose sides are geodesics.

MORE THAN ONE PARALLEL GEODESIC

Now consider a surface in which there can be two or more geodesics through a point, parallel to a given geodesic. This form of non-Euclidean geometry is known as *hyperbolic geometry*. It is also called *Lobachevskian geometry*, named after Nikolai Lobachevsky, a Russian mathematician who lived from 1793 until 1856.

A Lobachevskian universe is said to have *negative curvature*. A surface with negative curvature is warped in different senses, depending on how the axis is oriented. Examples of 2D universes with negative curvature are extended saddle-shaped and funnel-shaped surfaces.

Figure 11-10 is an illustration of a negatively curved surface containing a triangle and a quadrilateral. On this surface, the interior angles of the triangle and the quadrilateral add up to less than 180° and 360°, respectively. The measures of the interior angles of a polygon on a Lobachevskian surface always sum up to less than the sum of the measures of the interior angles of a similar polygon on a flat plane.

Curved Space

The observable universe seems, upon casual observation, to be Euclidean. If you use lasers to "construct" polygons and then measure their interior angles

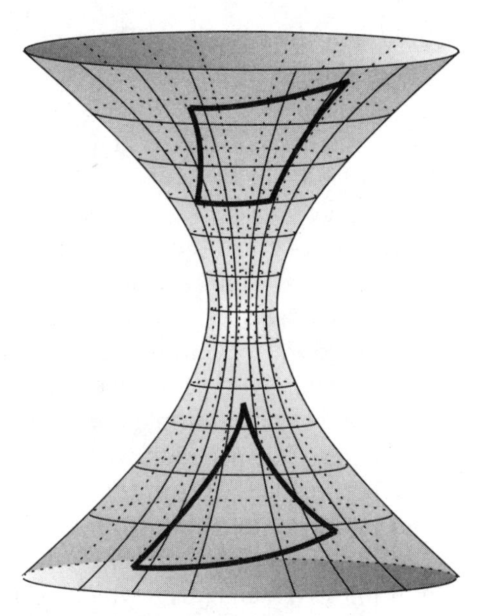

Fig. 11-10. An example of a surface with negative curvature, showing a triangle and a quadrilateral whose sides are geodesics.

with precision lab equipment, you'll find that the angle measures add up according to the rules of Euclidean geometry. The conventional formulas for the volumes of solids such as the pyramid, cube, and sphere hold perfectly, as far as anyone can tell. Imagine a 3D space in which these rules do not hold! This is called *curved 3D space*, *warped 3D space*, or *non-Euclidean 3D space*. It is the 3D analog of a non-Euclidean 2D surface.

GRAVITY WARPS SPACE

There is evidence that the 3D space in which we live is not perfectly Euclidean. Gravitational fields produce effects on light beams that suggest a Lobachevskian sort of warping—a negative curvature—of 3D space. Under ordinary circumstances this warping is so subtle that we don't notice it, but it has been detected by astronomers using sensitive equipment, and in exceptional cases it can be directly observed.

The behavior of light from distant stars has been carefully observed as the rays pass close to the sun during solar eclipses. The idea is to find out whether or not the sun's gravitational field, which is strong near the surface, bends light rays in the way that we should expect if space has negative curvature. Early in the 20th century, Albert Einstein predicted that such bending could be observed and measured, and he calculated the expected angular changes that should be seen in the positions of distant stars as the sun passes almost directly in front of them. Repeated observations have shown Einstein to be correct, not only as to the existence of the spatial curvature, but also to its extent as a function of distance from the sun. As the distance from the sun increases, the spatial warping decreases. The greatest amount of light-beam bending occurs when the photons graze the sun's surface.

In another experiment, the light from a distant, brilliant object called a *quasar* is observed as it passes close to a compact, dark mass that astronomers think is an intense source of gravitation known as a *black hole*. The light-bending is much greater near this type of object than is the case near the sun. The rays are bent enough so that multiple images of the quasar appear, with the black hole at the center. One peculiar example, in which four images of the quasar appear, has been called a *gravitational light cross*.

Any source of gravitation, no matter how strong or weak, is attended by curvature of the 3D space in its vicinity, such that light rays follow geodesic paths that are not straight lines. Which causes which? It is a chicken-and-egg mystery. Does spatial curvature cause gravitation, or do gravitational fields cause warping of space? Are both effects the result of some other phenomenon that has yet to be defined and understood? Such questions are of interest to astronomers and cosmologists. For the mathematician, it is enough to know that the curvature exists and can be defined. It's more than a product of someone's imagination.

THE "HYPERFUNNEL"

The curvature of space in the presence of a strong gravitational field has been likened to a funnel shape (Fig. 11-11), except that the surface of the funnel is 3D rather than 2D, and the entire object is 4D rather than 3D. When the fourth dimension is defined as time, the mathematical result is that time flows more slowly in a gravitational field than it does in interstellar or intergalactic space. This has been experimentally observed.

The shortest distance in 3D space between any two points near a gravitational source is a geodesic arc, not a straight line segment. Curvature of space caused by gravitational fields always increases the distances between points in the vicinity of the source of the gravitation. The shortest path between any two points in non-Euclidean space is always greater than the distance would be if the space between the points were Euclidean. As the intensity of the gravitation increases, the extent of the spatial curvature also increases. There is some effect, theoretically, even when gravity is weak.

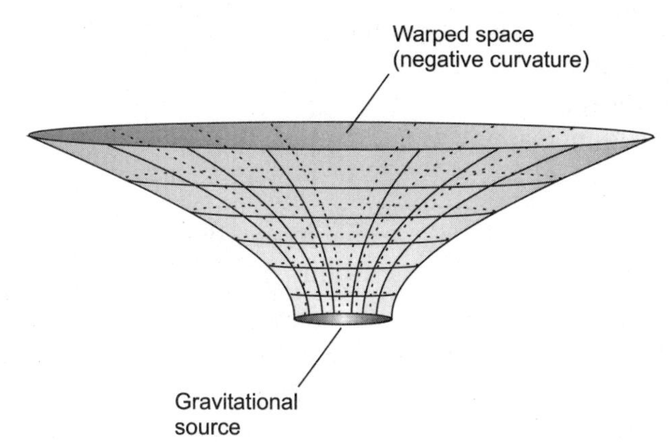

Fig. 11-11. An intense source of gravitation produces negative curvature, or warping, of space in its immediate vicinity.

Any particle that has mass is surrounded by its own gravitational field, so the earth has its own shallow "gravitational hyperfunnel." So does the moon; so do asteroids; so do meteoroids and space dust particles. What about the shape of space on an intergalactic scale? Does the entire universe, containing all the stars, galaxies, quasars, and other stuff that exists, have a geometric shape? If so, is it Riemannian, Lobachevskian, or flat?

Quiz

Refer to the text in this chapter if necessary. A good score is eight correct. Answers are in the back of the book.

1. Suppose there are two points P and Q in deep space, separated by exactly 1 second-equivalent (or 1 light-second). At point P there is a space traveler with a powerful pulsed laser. At point Q there is a mirror oriented so the space traveler can shine the laser at it and see the reflected pulse exactly 2 seconds later. Suppose an extremely dense, dark neutron star passes almost (but not quite) between points P and Q, producing a pronounced negative curvature of space in the region. If the space traveler shines the laser toward point Q while the neutron star is near the line of sight, the return pulse will
   (a) be seen after exactly 2 seconds
   (b) be seen after a little more than 2 seconds
   (c) be seen after a little less than 2 seconds
   (d) never be seen

2. If there are no pairs of parallel geodesics on a surface, that surface is
   (a) Lobachevskian
   (b) Euclidean
   (c) four-dimensional
   (d) Riemannian

3. Einstein's "four-sphere" cosmic model describes a universe with
   (a) finite volume and finite radius
   (b) finite volume and infinite radius
   (c) infinite volume and finite radius
   (d) infinite volume and infinite radius

4. Considered with respect to the speed of light, one second-equivalent represents a distance of about
   (a) 300,000 meters
   (b) 300,000,000 meters
   (c) 300,000,000,000 meters
   (d) 300,000,000,000,000 meters

5. If the universe is portrayed as Cartesian four-space with three spatial dimensions and one time dimension, then the 4D path of a stationary point that "lasts forever" is
   (a) a straight line
   (b) a circle
   (c) a helix
   (d) a spiral

6. How many line-segment edges does a rectangular four-prism have?
   (a) 16
   (b) 24
   (c) 32
   (d) 48

7. The distance between (0,0,0,0,0,0) and (1,1,1,1,1,1) in Cartesian six-space is
   (a) equal to 1
   (b) equal to the square root of 2
   (c) equal to the sixth root of 2
   (d) equal to the square root of 6

8. Suppose a light bulb is switched on, and the photons travel outward in ever-expanding spherical paths, as depicted in the dimensionally reduced drawing of Fig. 11-6. In this graph, the flare angle of the

cone (the angle between the positive t axis and any ray extending from the origin outward along the cone's surface) is
(a) 30°
(b) 45°
(c) 90°
(d) impossible to determine without more information

9. The 4D hypervolume of a rectangular four-prism consisting of a cube measuring 200,000 kilometers on each edge, and whose life is represented by a single point in time, is equal to
(a) zero
(b) 2/3 of a second-equivalent
(c) 4/9 of a second-equivalent
(d) 8/27 of a second-equivalent

10. Suppose P and Q are points on a surface that is positively curved. The length of the line segment PQ in 3D space
(a) is the same as the length of the geodesic PQ
(b) is greater than the length of the geodesic PQ
(c) is less than the length of the geodesic PQ
(d) might be equal to, less than, or greater than the length of the geodesic PQ, depending on the locations of P and Q

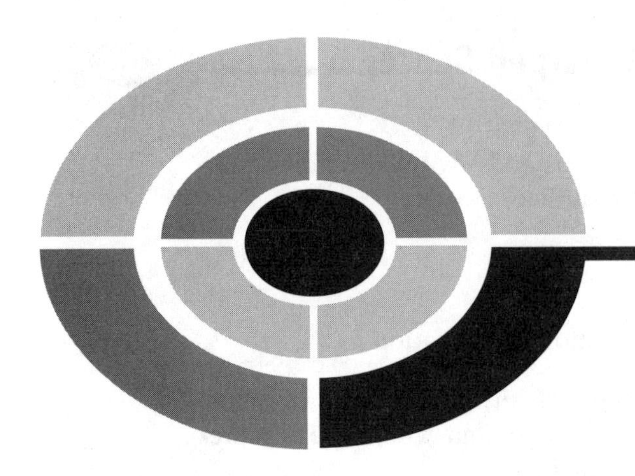

Test: Part Two

Do not refer to the text when taking this test. You may draw diagrams or use a calculator if necessary. A good score is at least 38 correct. Answers are in the back of the book. It's best to have a friend check your score the first time, so you won't memorize the answers if you want to take the test again.

1. Two straight lines that are not parallel, and that do not lie in the same plane, are called
   (a) orthogonal
   (b) perpendicular
   (c) non-Euclidean
   (d) normal
   (e) skew

2. The equation of the unit circle in polar coordinates is
   (a) $r = \theta$
   (b) $r = 1$
   (c) $r = 0$
   (d) $\theta = 1$
   (e) $\theta = 0$

3. The tangent function $y = \tan x$ is defined for all the following values of x except:

(a) $x = 0°$
(b) $x = 45°$
(c) $x = 90°$
(d) $x = 135°$
(e) $x = 180°$

4. Suppose a vector **a** begins at the point (2,3) and ends at the point (−5,6). In standard form, this vector is
(a) **a** = (−3,9)
(b) **a** = (7,−3)
(c) **a** = (−10,18)
(d) **a** = (0,0)
(e) none of the above

5. Refer to Fig. Test 2-1. Suppose that all the flat faces of the 3D figure are parallelograms, but none of the angles x, y, and z are right angles. If this is the case, we can nevertheless truthfully say that the figure is
(a) a parallelogram
(b) a rhombus
(c) a rectangular prism
(d) a parallelepiped
(e) none of the above

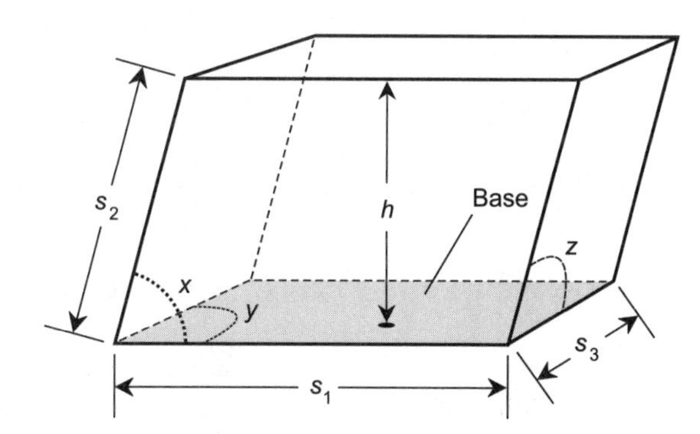

Fig. Test 2-1. Illustration for Questions 5, 6, and 7 in the test for Part Two.

6. Refer to Fig. Test 2-1. Imagine that the lengths of the edges s_1, s_2, and s_3 remain constant. Suppose the angles x, y, and z are all right angles at first, but they uniformly and gradually decrease as the 3D object is "squashed down." What happens to the volume of the 3D object?
(a) It increases
(b) It remains constant

(c) It decreases
(d) Nothing, because the scenario described is impossible
(e) We can't say without more information

7. Refer to Fig. Test 2-1. The overall surface area, A, of the 3D object shown is given by the following equation:

$$A = 2s_1s_2 \sin x + 2s_1s_3 \sin y + 2s_2s_3 \sin z$$

where sin x represents the sine of angle x, sin y represents the sine of angle y, and sin z represents the sine of angle z. Suppose we start with a situation where x, y, and z are all right angles. Then, without changing the lengths s_1, s_2, or s_3, we "squash" the object so all three angles x, y, and z measure 30°. What happens to the overall surface area of the object?
(a) It does not change
(b) It decreases by a factor of 2
(c) It decreases by a factor of 4
(d) It decreases by a factor of 8
(e) We can't say without more information

8. The graph of $4r = 3\theta$ in the mathematician's polar coordinate system looks like
(a) a spiral
(b) a cardioid
(c) a circle
(d) a three-leafed rose
(e) a four-leafed rose

9. The distance between (0,0,0,0,0) and (1,1,1,1,1) in Cartesian five-space is
(a) equal to 1
(b) equal to the square root of 2
(c) equal to the fifth root of 2
(d) equal to the square root of 5
(e) equal to the fifth root of 5

10. In order to uniquely define a point in 4D time-space
(a) four coordinates in time must be identified
(b) four coordinates in space must be identified
(c) three coordinates in space and one coordinate in time must be specified

(d) three coordinates in time and one coordinate in space must be specified
(e) two coordinates in time and two coordinates in space must be specified

11. Suppose two non-perpendicular planes intersect. The angle at which the planes intersect can be defined in two ways: as an acute angle u or as an obtuse angle v. If u and v are expressed in radians, then
(a) $u + v = 1$
(b) $u + v = \pi/2$
(c) $u + v = \pi$
(d) $u + v = 2\pi$
(e) none of the above

12. Refer to Fig. Test 2-2. The coefficients a, b, and c of vector $\mathbf{m} = (a,b,c)$, which is normal (perpendicular) to plane W, are sufficient to uniquely determine
(a) the equation of plane W
(b) the orientation of plane W
(c) whether or not plane W passes through the origin
(d) all of the above (a), (b), and (c)
(e) none of the above (a), (b), or (c)

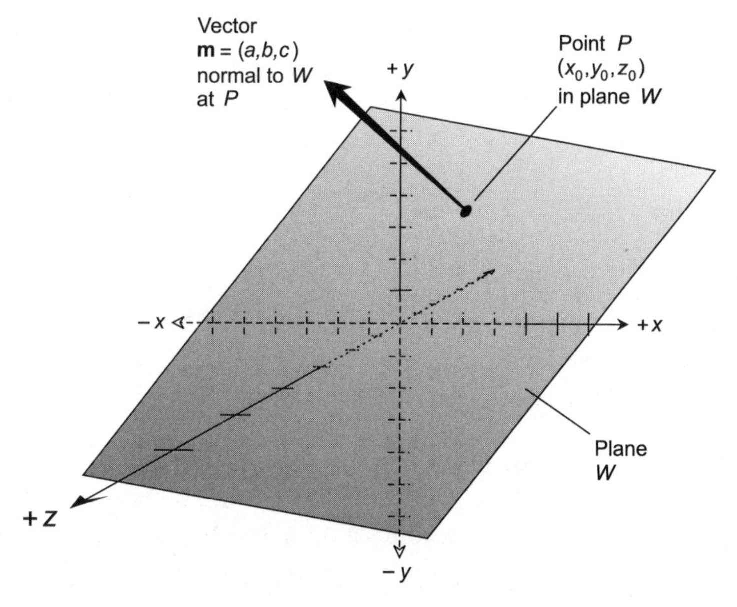

Fig. Test 2-2. Illustration for Questions 12, 13, and 14 in the test for Part Two.

13. Refer to Fig. Test 2-2. Suppose we know the coefficients a, b, and c of vector $\mathbf{m} = (a,b,c)$, which is normal to plane W. Suppose we also know the coordinates (x_0,y_0,z_0) of point P. This information is sufficient to uniquely determine
    (a) the equation of plane W
    (b) the orientation of plane W
    (c) whether or not plane W passes through the origin
    (d) all of the above (a), (b), and (c)
    (e) none of the above (a), (b), or (c)

14. Refer to Fig. Test 2-2. Suppose we know the coefficients a, b, and c of vector $\mathbf{m} = (a,b,c)$, which is normal to plane W. Suppose we also know the coordinates (x_0,y_0,z_0) of point P. Now, imagine that we multiply all the values a, b, and c by -1, obtaining the vector $-\mathbf{m} = (-a,-b,-c)$. Let the resulting plane, which is determined by the point P and which is normal to the vector $-\mathbf{m}$, be called plane X. Which of the following is true?
    (a) Planes X and W are perpendicular
    (b) Planes X and W are distinct, but parallel
    (c) Planes X and W coincide
    (d) Plane X cannot be defined because there are infinitely many possibilities
    (e) Plane X cannot exist

15. Suppose two planes intersect at an angle of 140°. This is the less common of two ways the intersection angle can be expressed. The more common value for the intersection angle of these two planes is
    (a) 40°
    (b) –40°
    (c) –140°
    (d) 220°
    (e) none of the above

16. The dot product of the vectors $\mathbf{a} = (2,4,-1)$ and $\mathbf{b} = (-5,1,2)$ in Cartesian xyz-space is equal to
    (a) the scalar quantity –8
    (b) the vector $(-10,4,-2)$
    (c) the scalar quantity 80
    (d) a vector perpendicular to the plane containing **a** and **b**
    (e) a vector in the plane containing **a** and **b**

17. Suppose the spherical coordinates of a certain object in the sky are specified as (θ,ϕ,r), where θ is its elevation with respect to the plane of

the horizon, ϕ is its azimuth, and r is its radius (distance from us). Imagine that the object flies horizontally away from us, traveling at the same heading as its azimuth as we see it (so ϕ remains constant). What happens to θ and r?

(a) The value of θ approaches 0, while r increases without limit
(b) The values of θ and r both approach 0
(c) The value of θ increases without limit, while r approaches 0
(d) The values of θ and r both increase without limit
(e) None of the above

18. Suppose the spherical coordinates of a certain object in the sky are specified as (θ,ϕ,r), where θ is its elevation with respect to the plane of the horizon, ϕ is its azimuth, and r is its radius (distance from us). Imagine that the object flies straight away from us. What happens to θ and r?
    (a) The value of θ approaches 0, while r increases without limit
    (b) The values of θ and r both approach 0
    (c) The value of θ increases without limit, while r approaches 0
    (d) The values of θ and r both increase without limit
    (e) None of the above

19. Considered with respect to the speed of light, one minute-equivalent represents a distance of about
    (a) 18,000 kilometers
    (b) 180,000 kilometers
    (c) 1,800,000 kilometers
    (d) 18,000,000 kilometers
    (e) 180,000,000 kilometers

20. The faces (including the base) of a rectangular pyramid are all
    (a) triangles
    (b) squares
    (c) rectangles
    (d) rhombuses
    (e) plane polygons

21. Suppose there are two planes X and Y such that, for all lines L passing through both X and Y, the acute angle between L and X has the same measure as the acute angle between L and Y. From this information, it is reasonable to suppose that
    (a) planes X and Y are perpendicular
    (b) planes X and Y are non-Euclidean
    (c) planes X and Y are parallel

(d) there are no lines parallel to either plane X or plane Y
(e) there are no lines perpendicular to either plane X or plane Y

22. How many straight line-segment edges does a tesseract (or four-cube) have?
(a) 16
(b) 24
(c) 32
(d) 48
(e) 96

23. What is the 4D hypervolume of a tesseract (or four-cube) measuring 1 meter-equivalent on each edge? (Call the standard unit of 4D hypervolume a *quartic meter-equivalent*.)
(a) 1 quartic meter-equivalent
(b) 4 quartic meter-equivalents
(c) 16 quartic meter-equivalents
(d) 64 quartic meter-equivalents
(e) It is impossible to say without more information

24. What is the 5D hypervolume of a five-cube measuring 1 meter-equivalent on each edge? (Call the standard unit of 5D hypervolume a *quintic meter-equivalent*.)
(a) 1 quintic meter-equivalent
(b) 5 quintic meter-equivalents
(c) 25 quintic meter-equivalents
(d) 125 quintic meter-equivalents
(e) It is impossible to say without more information

25. Given a slant circular cone whose base has radius r and whose height, expressed between the apex and the plane containing the base, is h, the volume V is given by the following formula:

$$V = \pi r^2 h/3$$

Let K_1 and K_2 be slant circular cones. Suppose the radius of K_2 is twice as great as the radius of K_1, but the height of K_2 is only half the height of K_1. Which of the following statements is true?
(a) The volume of K_2 is four times the volume of K_1
(b) The volume of K_2 is twice the volume of K_1
(c) The volume of K_1 is four times the volume of K_2
(d) The volume of K_1 is twice the volume of K_2
(e) The volumes of K_1 and K_2 are the same

26. Refer to Fig. Test 2-3. If the rectangular coordinates x_0 and y_0 of point P are both doubled, what happens to the value of r_0?
    (a) It increases by a factor of the square root of 2
    (b) It doubles
    (c) It quadruples
    (d) It does not change
    (e) This question cannot be answered without more information

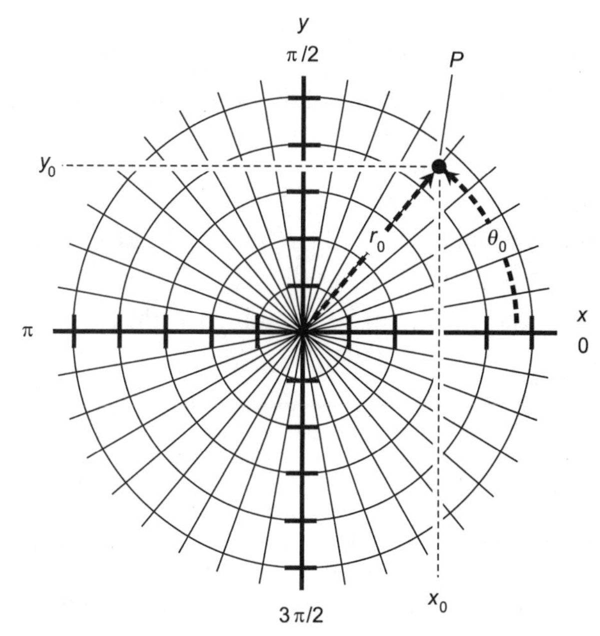

Fig. Test 2-3. Illustration for Questions 26, 27, and 28 in the test for Part Two.

27. Refer to Fig. Test 2-3. If the rectangular coordinates x_0 and y_0 of point P are both doubled, what happens to the value of θ_0?
    (a) It increases by a factor of the square root of 2
    (b) It doubles
    (c) It is multiplied by –1
    (d) It does not change
    (e) It increases by π rad

28. Refer to Fig. Test 2-3. If the rectangular coordinates x_0 and y_0 of point P are both multiplied by –1, what happens to the value of θ_0?
    (a) It increases by a factor of the square root of 2
    (b) It doubles
    (c) It is multiplied by –1

(d) It does not change
(e) It increases by π rad

29. What is the volume of a rectangular prism that is 200 millimeters high, 700 millimeters wide, and 500 millimeters deep?
(a) 1.4 square meters
(b) 1.4 cubic meters
(c) 0.07 square meters
(d) 0.07 cubic meters
(e) None of the above

30. A donut-shaped geometric solid is called a
(a) right circular cylinder
(b) frustum of a cylinder
(c) bent circular cylinder
(d) truncated ellipsoid
(e) none of the above

31. A single, specific line can be contained within
(a) one and only one plane
(b) parallel planes
(c) at most two planes
(d) at most three planes
(e) infinitely many planes

32. The entire polar coordinate plane, showing all possible points with angular values from 0° to 360° and radial values corresponding to any non-negative real number, can be portrayed within a finite circular region by
(a) using a logarithmic angular scale
(b) using a logarithmic radial scale
(c) geometric compression of the angular scale
(d) geometric compression of the radial scale
(e) no known means

33. The height, or altitude, in cylindrical coordinates is expressed
(a) in linear units, perpendicular to the plane in which the direction angle is expressed
(b) in linear units, in the plane in which the direction angle is expressed
(c) in linear units, in a direction parallel to the direction in which the radius is expressed.
(d) in degrees, as an angle relative to the horizon
(e) in radians, as an angle relative to the horizon

34. The direction angle in the mathematician's polar coordinate system is expressed
 (a) in a clockwise sense
 (b) in a counterclockwise sense
 (c) in either sense
 (d) only in radians
 (e) only in degrees

35. An example of a negatively curved 2D surface is
 (a) the surface of a sphere
 (b) the surface of an ellipsoid
 (c) the surface of a tesseract
 (d) the surface of a four-sphere
 (e) none of the above

36. The product of the vectors **a** = (–3,0,4) and **b** = (2,1,–5) in Cartesian *xyz*-space is equal to
 (a) the scalar quantity –26
 (b) the scalar quantity –1
 (c) the vector (–6,0,–20)
 (d) the vector (–1,1,–1)
 (e) this problem cannot be solved without more information

37. The sum of the vectors **a** = (–3,0,4) and **b** = (2,1,–5) in Cartesian *xyz*-space is equal to
 (a) the scalar quantity –26
 (b) the scalar quantity –1
 (c) the vector (–6,0,–20)
 (d) the vector (–1,1,–1)
 (e) this problem cannot be solved without more information

38. Given a slant circular cylinder whose base has radius r and whose height, expressed between the top and the plane containing the base, is h, the volume V is given by the following formula:

$$V = \pi r^2 h$$

Let C_1 and C_2 be slant circular cylinders. Suppose the radius of C_2 is four times as great as the radius of C_1, but the height of C_2 is only 1/16 the height of C_1. Which of the following statements is true?

 (a) The volume of C_2 is four times the volume of C_1
 (b) The volume of C_2 is twice the volume of C_1
 (c) The volume of C_1 is four times the volume of C_2

(d) The volume of C_1 is twice the volume of C_2
(e) The volumes of C_1 and C_2 are the same

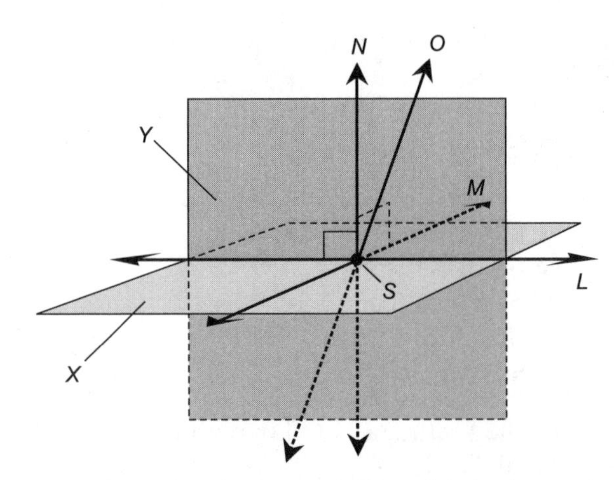

Fig. Test 2-4. Illustration for Questions 39 and 40 in the test for Part Two.

39. Let X be a plane. Suppose a line O, which is not normal to plane X, intersects plane X at some point S as shown in Fig. Test 2-4. Let N be a line normal to plane X, passing through point S. Let Y be the plane determined by the intersecting lines N and O. Let L be the line formed by the intersection of planes X and Y. Let M be a line in plane X that passes through point S, but is different from line L. The angle between line O and plane X is the same as
   (a) the angle between line O and line L
   (b) the angle between line O and line M
   (c) the angle between line O and plane Y
   (d) the angle between line L and line M
   (e) the angle between line L and line N

40. Examine Fig. Test 2-4. Suppose lines L, M, and N are all mutually perpendicular, like the axes of Cartesian xyz-space, and they all intersect at point S. Also suppose line L is common to both planes X and Y. If line M is in plane X and line N is in plane Y, then
   (a) planes X and Y are parallel
   (b) planes X and Y are obtuse
   (c) planes X and Y are skew
   (d) planes X and Y are acute
   (e) planes X and Y are perpendicular

41. Two intersecting lines define a single
    (a) line
    (b) ray
    (c) plane
    (d) triangle
    (e) quadrilateral

42. The faces of a parallelepiped are all
    (a) triangles
    (b) squares
    (c) rectangles
    (d) rhombuses
    (e) parallelograms

43. If a straight line in Cartesian *xyz*-space has direction defined by $\mathbf{m} = 2\mathbf{i} + 2\mathbf{j} + 2\mathbf{k}$, we can surmise
    (a) that the line is parallel to the x axis
    (b) that the line is parallel to the y axis
    (c) that the line is parallel to the z axis
    (d) that the line is not parallel to the x axis, the y axis, or the z axis
    (e) that the line passes through the origin

44. The smaller of the two definable angles between a line and a plane has a measure that can range anywhere between 0 rad and
    (a) $\pi/4$ rad
    (b) $\pi/2$ rad
    (c) π rad
    (d) $3\pi/2$ rad
    (e) 2π rad

45. Imagine two vectors in Cartesian *xyz*-space. Suppose vector **a** begins at the origin and ends at the point $(x_a,y_a,z_a) = (2,3,5)$. Suppose vector **b** begins at point (1,1,1) and ends at the point $(x_b,y_b,z_b) = (3,4,6)$. What can we say about these two vectors?
    (a) They are parallel to each other, but they are not equivalent
    (b) They are equivalent, but they are not parallel to each other
    (c) They are equivalent, and they are parallel to each other
    (d) They are parallel to each other, but they point in opposite directions
    (e) They are skewed with respect to each other

46. A significant difference between the mathematician's polar coordinate plane and the navigator's polar coordinate plane is
    (a) that the radii are measured in opposite directions

(b) that the direction angles are expressed in opposite senses
(c) that one scheme uses linear units, and the other does not
(d) that the radii are expressed in degrees in one scheme, and in radians in the other scheme
(e) nothing; there is no difference between the two systems

47. In the dimensionally reduced illustration Fig. Test 2-5 showing the earth's path through time-space, each vertical division represents $^1/_4$ of a year, or approximately 91.3 days. Suppose the vertical scale is changed so that the pitch of the helix becomes only half as great (as if it were a spring compressed by a factor of 2). Further suppose that the horizontal scales remain unchanged. Then each vertical division represents
(a) approximately 22.8 days
(b) approximately 45.7 days
(c) approximately 91.3 days
(d) approximately 183 days
(e) approximately 365 days

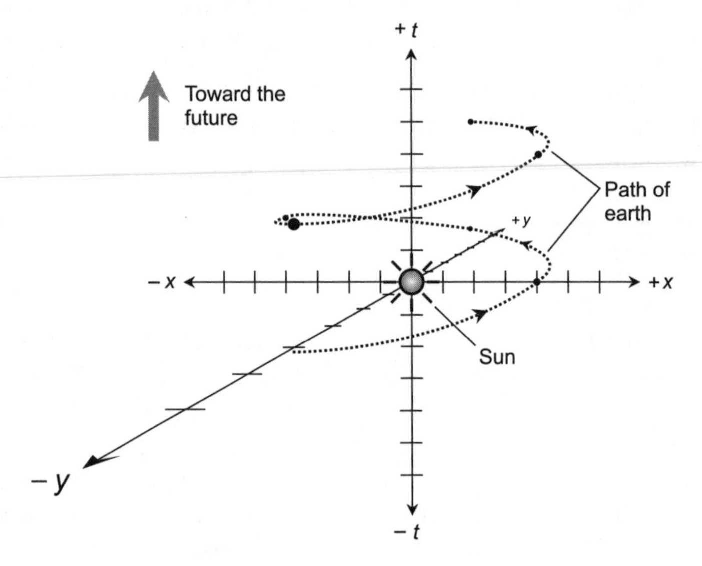

Fig. Test 2-5. Illustration for Questions 47 and 48 in the test for Part Two.

48. In the dimensionally reduced illustration Fig. Test 2-5 showing the earth's path through time-space, each vertical division represents $^1/_4$ of a year, or approximately 91.3 days. Suppose the x and y scales are both changed so that the radius of the helix becomes twice as great.

Further suppose that the vertical scale remains unchanged. Then each vertical division represents
(a) approximately 22.8 days
(b) approximately 45.7 days
(c) approximately 91.3 days
(d) approximately 183 days
(e) approximately 365 days

49. Suppose there are two ellipsoids E_1 and E_2 that have identical proportions, but the radii of E_2 are all exactly four times the radii of E_1. Suppose V_1 is the volume of E_1, and V_2 is the volume of E_2. Which, if any, of the following equations (a), (b), (c), or (d) is true?
(a) $V_2 = 4V_1$
(b) $V_2 = 8V_1$
(c) $V_2 = 16V_1$
(d) $V_2 = 32V_1$
(e) None of the above equations (a), (b), (c), or (d) is true

50. The distance between a point and a plane is expressed along
(a) a line passing through the point and parallel to the plane
(b) a line passing through the point and skewed relative to the plane
(c) a line passing through the point and normal to the plane
(d) a line passing through the point and contained in the plane
(e) none of the above

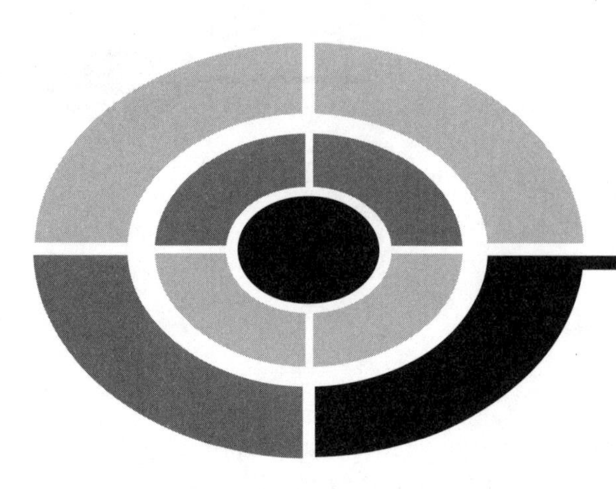

Final Exam

Do not refer to the text when taking this test. You may draw diagrams or use a calculator if necessary. A good score is at least 75 correct. Answers are in the back of the book. It's best to have a friend check your score the first time, so you won't memorize the answers if you want to take the test again.

1. Imagine two triangles such that their corresponding sides have equal lengths as you proceed around them both in the same direction. These two triangles are
   (a) isosceles
   (b) non-Euclidean
   (c) equilateral
   (d) directly congruent
   (e) symmetrical

2. Suppose a rhombus has diagonals of equal length. From this we can conclude that the rhombus is
   (a) a rectangle
   (b) irregular
   (c) a square
   (d) a trapezoid
   (e) none of the above

3. Look at Fig. Exam-1. Which, if any, of the following statements (a), (b), (c), or (d) is true?
   (a) $\triangle PSR$ is an equilateral triangle
   (b) Line segments PS and SR are equally long
   (c) $\triangle PSR$ is congruent to $\triangle PSQ$
   (d) $\angle QPS = \angle QSP$
   (e) None of the above statements is true

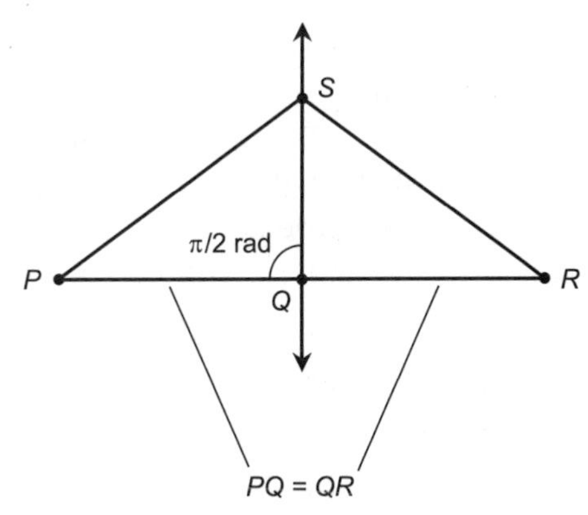

Fig. Exam-1. Illustration for Questions 3, 4, and 5 in the final exam.

4. In Fig. Exam-1, the measure of $\angle RQS$ is
   (a) equal to 2π rad minus the sum of the measures of $\angle PSQ$ and $\angle RSQ$
   (b) equal to 180° minus the measure of $\angle PQS$
   (c) equal to twice the measure of $\angle PSQ$
   (d) equal to half the sum of the measures of $\angle PSQ$ and $\angle RSQ$
   (e) not equal to anything described above

5. Fill in the blank: In Fig. Exam-1, line QS is _______ line segment PR.
   (a) parallel to
   (b) congruent to
   (c) a perpendicular bisector of
   (d) a parallel bisector of
   (e) in a different plane than

6. The formula for the interior area, A (in square units) of a regular polygon with n sides of length s is:

$$A = (ns^2/4) \cot (180°/n)$$

What is the interior area of a regular heptagon (7-sided polygon) with sides each measuring 1.000 meter in length? You may use a calculator to determine this. Express the answer to two decimal places. The cotangent (cot) of an angle is equal to the cosine (cos) divided by the sine (sin).

(a) 0.84 square meters
(b) 3.63 square meters
(c) 5.32 square meters
(d) 8.33 square meters
(e) 10.00 square meters

7. Refer to Fig. Exam-2. If $r_1 = r_2$, then
(a) the interior area of the figure is equal to πr_2^2
(b) the interior area of the figure is equal to $2r_2^2$
(c) the perimeter of the figure is equal to $2r_1$
(d) the perimeter of the figure is equal to πr_2
(e) none of the above

8. In Fig. Exam-2, the ratio of r_1, the length of the major semi-axis, to r_2, the length of the minor semi-axis, is called the
(a) elongation
(b) eccentricity
(c) ellipticity
(d) deviation
(e) oblongation

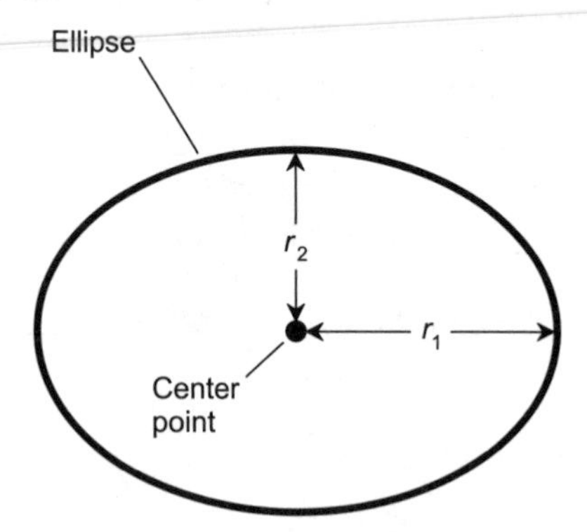

Fig. Exam-2. Illustration for Questions 7, 8, and 9 in the final exam.

9. Refer to Fig. Exam-2. The interior area, A, of the ellipse is given by:

$$A = \pi r_1 r_2$$

Based on this formula, if the length of the major semi-axis of an ellipse is tripled, what must happen to the length of the minor semi-axis in order for the interior area to remain the same?

(a) We cannot say, because the formula does not contain enough information
(b) The minor semi-axis must become 1/3 as great
(c) The minor semi-axis must become 1/6 as great
(d) The minor semi-axis must become 1/9 as great
(e) The minor semi-axis must become 1/27 as great

10. In Euclidean plane geometry, how many points are required to uniquely define a single straight line?
(a) None
(b) One
(c) Two
(d) Three
(e) Four

11. Imagine that you have a telescope equipped with a camera. You focus on a distant, triangular sign and take a photograph of it. Then you triple the magnification of the telescope and, making sure the whole sign fits into the field of view of the camera, you take another photograph. When you get the photos developed, you see triangles in each photograph. No matter what else might be true about this scenario, we can conclude for certain that the two triangles in the photographs must be
(a) equilateral
(b) symmetrical
(c) non-Euclidean
(d) isosceles
(e) none of the above

12. A rhombus is a geometric figure in which
(a) all the sides are equally long
(b) all the angles have the same measure
(c) the sum of the measures of the interior angles is 720°
(d) no two points lie in the same plane
(e) at least one of the sides is infinitely long

13. Refer to Fig. Exam-3. What is the equation of line L?
(a) $-x + y - 5 = 0$
(b) $-x + y + 5 = 0$
(c) $x + y + 5 = 0$
(d) $-2x + 3y - 1 = 0$
(e) $2x - 3y + 1 = 0$

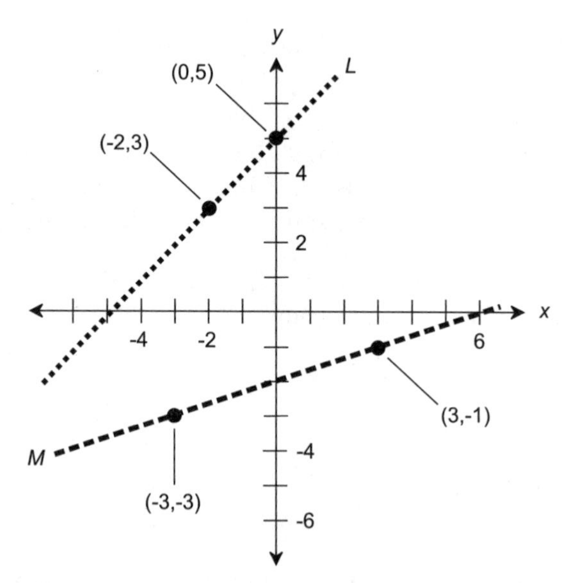

Fig. Exam-3. Illustration for Questions 13, 14, and 15 in the final exam.

14. Refer to Fig. Exam-3. What is the equation of line M?
    (a) $3x - y = 0$
    (b) $x - 3y + 6 = 0$
    (c) $x + 3y + 6 = 0$
    (d) $-x + 3y + 6 = 0$
    (e) $-3x - 3y = 0$

15. Refer to Fig. Exam-3. Imagine curves L and M as infinitely long, straight lines. Do these lines intersect? If so, what are the coordinates (x_0,y_0) of their intersection point?
    (a) The lines intersect at $(x_0,y_0) = (-21,-11)$
    (b) The lines intersect at $(x_0,y_0) = (-21/2,-11/2)$
    (c) The lines intersect at $(x_0,y_0) = (-10,-5)$
    (d) The lines intersect, but more information is needed to figure out where
    (e) The lines do not intersect

16. What is the distance r between the points $(-3,-3)$ and $(-6,-7)$ on the Cartesian plane?
    (a) $r = 3$
    (b) $r = 4$
    (c) $r = 5$
    (d) $r = (-9,-10)$
    (e) $r = (3,4)$

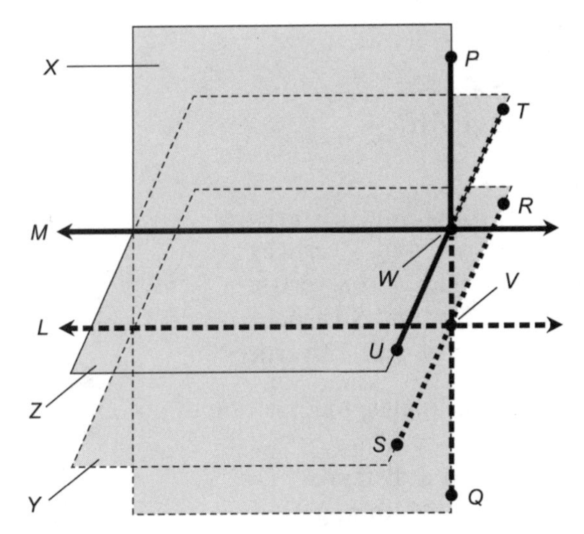

Fig. Exam-4. Illustration for Questions 17 and 18 in the final exam.

17. Examine Fig. Exam-4. Suppose planes *Y* and *Z* are parallel, and they intersect plane *X* in lines *L* and *M*, respectively. Suppose line *PQ* is in plane *X*, and is perpendicular to both lines *L* and *M*. Suppose line *RS* is in plane *Y*, and is perpendicular to line *L*. Suppose line *TU* is in plane *Z*, and is perpendicular to line *M*. Let *V* and *W* be intersection points among the lines, as shown. From these facts, we can conclude that
    (a) $\angle PWT = \angle PVS$
    (b) $\angle PWT = \angle QVR$
    (c) $\angle PWT = \angle QVS$
    (d) $\angle PWT = \angle QWT$
    (e) $\angle PWT = \angle UWP$

18. In Fig. Exam-4, assume all of the conditions described in Question 17 hold true. Which of the following are corresponding angles?
    (a) $\angle PWT$ and $\angle WVR$
    (b) $\angle PWT$ and $\angle WVS$
    (c) $\angle QWU$ and $\angle QWT$
    (d) $\angle QWU$ and $\angle QVR$
    (e) $\angle QVS$ and $\angle PVR$

19. Vertical (that is, opposite) dihedral angles between two intersecting planes always have measures that
    (a) add up to 90°

(b) add up to 180°
(c) add up to 360°
(d) are the same
(e) none of the above

20. A cube has
(a) 6 faces, 12 edges, and 12 vertices
(b) 6 faces, 8 edges, and 8 vertices
(c) 6 faces, 12 edges, and 8 vertices
(d) 8 faces, 8 edges, and 8 vertices
(e) 8 faces, 12 edges, and 12 vertices

21. In cylindrical coordinates, the position of a point is defined according to
(a) two angles and a distance
(b) two distances and an angle
(c) three distances
(d) three angles
(e) none of the above

22. The equation $x^2 + y^2 = 1$ can be used to define
(a) the exponential function
(b) the logarithmic function
(c) the cosine function
(d) the hyperbolic function
(e) none of the above

23. The tangent function $y = \tan x$ is defined for all the following values of x except:
(a) $x = 0°$
(b) $x = 45°$
(c) $x = 90°$
(d) $x = 135°$
(e) $x = 180°$

24. Imagine a vector **a** that has magnitude of 3 and points straight up (elevation 90°), and a vector **b** that has magnitude 4 and points toward the western horizon (azimuth 270°, elevation 0°). The cross product of **a** and **b**, written **a** × **b**, is
(a) a scalar equal to 12
(b) a scalar equal to 0
(c) a vector with magnitude 12, pointing toward the west but above the horizon

(d) a vector with magnitude 12, pointing toward the southern horizon
(e) impossible to determine without more information

25. Imagine a vector **a** that has magnitude of 3 and points straight up (elevation 90°), and a vector **b** that has magnitude 4 and points toward the western horizon (azimuth 270°, elevation 0°). The dot product of **a** and **b**, written **a • b**, is
(a) a scalar equal to 12
(b) a scalar equal to 0
(c) a vector with magnitude 12, pointing toward the west but above the horizon
(d) a vector with magnitude 12, pointing toward the southern horizon.
(e) impossible to determine without more information

26. When using a drafting compass and straight edge to perform a geometric construction, you must
(a) not make use of any calibrated scales
(b) always use a pen, not a pencil, to make markings on the paper
(c) use both instruments at least once
(d) never draw circles of arbitrary radius
(e) never draw line segments of arbitrary length

27. A circle is a specific type of
(a) polygon with infinitely many sides
(b) ellipse
(c) cone
(d) parabola
(e) none of the above

28. Refer to Fig. Exam-5. The top and the base of the figure are circles, with center points P and Q, as shown. This object is called a
(a) trapezoidal cone
(b) trapezoidal cylinder
(c) frustum of a cone
(d) partial cone
(e) truncated cylinder

29. Let S be the slant surface area of the object in Fig. Exam-5, that is, the area not including the base or the top. Let s be the slant height, r_1 be the radius of the circular top, and r_2 be the radius of the circular base. Let P be the center of the circular top, and Q be the center of the

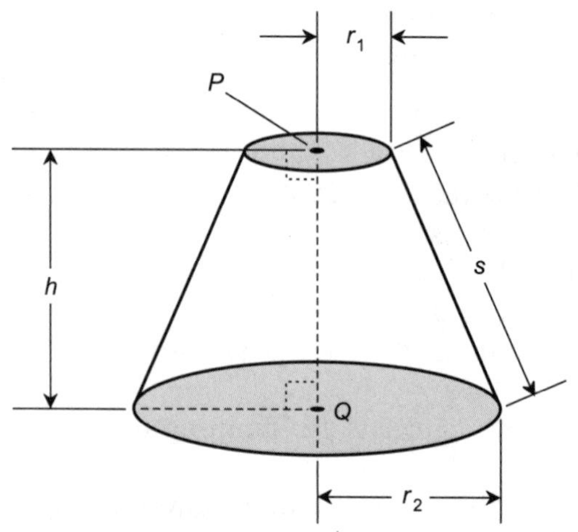

Fig. Exam-5. Illustration for Questions 28, 29, and 30 in the final exam.

circular base. Let h be the height of the figure. The value of S is given by the following formula:

$$S = \pi s(r_1 + r_2)$$

where π is approximately equal to 3.14159. What happens to the slant surface area if all the dimensions of this object are doubled?
(a) It does not change
(b) It doubles
(c) It becomes four times as great
(d) It becomes eight times as great
(e) It is impossible to say without more information

30. Refer to Fig. Exam-5. Suppose that $r_2 = 2r_1$. Imagine some plane X that contains the line segment connecting points P and Q. The intersection between plane X and the surface of the object, including the base and the top, is
(a) a triangle
(b) a rhombus
(c) a rectangle
(d) a parallelogram
(e) none of the above

31. Suppose a triangle has a base length of 10 meters and a height of 6 meters. The interior area of this triangle is
(a) approximately 7.75 square meters
(b) 20 square meters

(c) 30 square meters
(d) 60 square meters
(e) impossible to determine without more information

32. Suppose a triangle has a base length of 10 meters, and the other two sides both measure 8 meters in length. How long is the line segment that joins the midpoints of the sides that are 8 meters long?
(a) 2 meters
(b) 4 meters
(c) 5 meters
(d) 8 meters
(e) This question cannot be answered without more information

33. Suppose a triangle has a base length of 7 meters, and the other two sides both measure 11 meters in length. The line segment that joins the midpoints of the sides that are 11 meters long is
(a) perpendicular to those two sides
(b) parallel to those two sides
(c) normal to the base
(d) isosceles to the base
(e) parallel to the base

34. When two planes intersect, the measures of the adjacent dihedral angles defined by the intersection add up to
(a) 90°
(b) 180°
(c) 270°
(d) 360°
(e) 540°

35. Consider the mean radius of the earth to be 6371 kilometers. Consider the mean radius of the sun to be 696,000 kilometers. The volume of the sun is
(a) approximately 109 times the volume of the earth
(b) approximately 11,900 times the volume of the earth
(c) approximately 1,300,000 times the volume of the earth
(d) approximately 142,000,000 times the volume of the earth
(e) impossible to determine without more information

36. Imagine a torus T whose inner radius is half its outer radius. Suppose a line L passes through the center point of T. The intersection of L and T is
(a) four points

(b) three points
(c) two points
(d) the empty set (no points)
(e) impossible to determine without more information

37. A vector that begins at the origin of a coordinate system, and points outward in a specific direction (or orientation) from there, is said to be
(a) in unit form
(b) in zero form
(c) in real-number form
(d) in Euclidean form
(e) in standard form

38. Suppose we set off on a bearing of 90° in the navigator's polar coordinate system. We stay on a straight course. If the starting point is considered the origin, what is the graph of our path in Cartesian coordinates?
(a) $y = x$, where $x \leq 0$
(b) $y = 0$, where $x \geq 0$
(c) $x = 0$, where $y \geq 0$
(d) $y = x$, where $x \geq 0$
(e) None of the above

39. Skew lines
(a) are parallel, but they intersect each other
(b) are orthogonal, but they do not intersect each other
(c) are not parallel, and they lie in different planes
(d) are not parallel, but they lie in a single plane
(e) none of the above

40. If two lines intersect and are perpendicular, then they
(a) lie in different planes
(b) are parallel
(c) are geodesics on a sphere
(d) lie in the same plane
(e) have a common center point

41. Consider a circle C that is inscribed by a regular polygon S having n sides, and that is circumscribed by a regular polygon T, also having n sides. As n increases without limit, what happens to the perimeters of S and T?
(a) Their ratio approaches 1:1
(b) Their difference becomes greater and greater

(c) Their product approaches 1
(d) Their ratio approaches $1{:}\pi$
(e) Their ratio approaches $1{:}2\pi$

42. What is the slope of the line represented by the equation $y - 2 = 4(x + 5)$?
(a) –2
(b) 2
(c) 4
(d) 5
(e) 20

43. Right ascension is measured eastward from the position of the sun in the heavens on approximately
(a) March 21
(b) June 21
(c) September 21
(d) December 21
(e) the middle of the period of daylight

44. Right ascension, declination, and radius together comprise a scheme of
(a) polar coordinates
(b) Cartesian coordinates
(c) spherical coordinates
(d) cylindrical coordinates
(e) logarithmic coordinates

45. The 3D surface of a 4D sphere (or four-sphere) is an example of
(a) a 2D space
(b) a finite but unbounded 3D space
(c) an infinite but bounded 3D space
(d) a finite, bounded 3D space
(e) none of the above

46. What are the points, if any, at which the circle $x^2 + (y - 1)^2 = 1$ intersects the y axis?
(a) It is impossible to tell without more information
(b) (0,0) and (0,2)
(c) (0,0) and (0,–2)
(d) (0,0)
(e) There are none

47. A triangle on a surface can have interior angles, each of whose measure is between
    (a) 0 rad and 1 rad
    (b) 0 rad and $\pi/2$ rad
    (c) 0 rad and π rad
    (d) 0 rad and 2π rad
    (e) π rad and 2π rad

48. Each interior angle of a regular hexagon measures
    (a) $\pi/6$ rad
    (b) $\pi/3$ rad
    (c) $\pi/2$ rad
    (d) $2\pi/3$ rad
    (e) $3\pi/2$ rad

49. The time equivalent of 1000 kilometers, using the speed of light (300,000,000 meters per second) as a standard, is approximately
    (a) 333 second-equivalents
    (b) 33.3 second-equivalents
    (c) 3.33 second-equivalents
    (d) 0.333 second-equivalents
    (e) none of the above

50. Suppose there is a geodesic L on a surface S. Let P be some point near, but not on, the geodesic L. Suppose there exist infinitely many geodesics $M_1, M_2, M_3, \ldots$ on the surface S that pass through point P and do not intersect geodesic L. From this we know that
    (a) the surface S is non-Euclidean
    (b) the geodesic L is a circle
    (c) the geodesic L is an ellipse
    (d) the surface S is a sphere
    (e) the set of circumstances described is impossible

51. Imagine an infinitely long, stationary, straight line that has always existed, exists at this moment, and always will exist in the future. In 4D Euclidean time-space, the path of this line is
    (a) a line
    (b) a half-plane
    (c) a plane
    (d) a sphere
    (e) a tesseract

52. Suppose that two straight lines intersect, forming four angles. The two angles opposite each other are called
    (a) interior angles
    (b) supplementary angles
    (c) complementary angles
    (d) alternate angles
    (e) vertical angles

53. In Fig. Exam-6, suppose the measure of $\angle QPR$ is 50°. The other two interior angles have measures x and y. Which of the following statements can be made with certainty?
    (a) $x + y = 2\pi$ rad
    (b) $x + y = \pi$ rad
    (c) $x - y = \pi/2$ rad
    (d) $x - y = 50°$
    (e) $x + y = 130°$

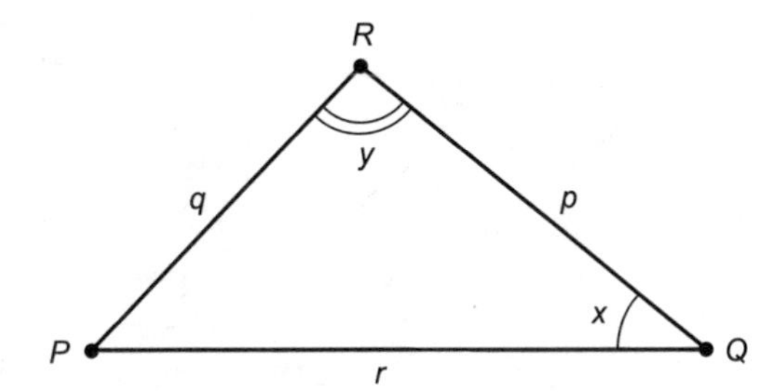

Fig. Exam-6. Illustration for Questions 53 and 54 in the final exam.

54. In Fig. Exam-6, suppose the measure of angle y is exactly 90°, the length of side r is exactly 10 meters, and the length of side p is exactly 8 meters. The length of side q is
    (a) approximately 1.732 meters
    (b) approximately 1.414 meters
    (c) exactly 6 meters
    (d) exactly 4 meters
    (e) impossible to determine without more information

55. Given any three distinct points, they cannot form a triangle if
    (a) each one is equidistant from the other two
    (b) they lie on different lines
    (c) they all lie on the surface of the same sphere
    (d) they all lie on the perimeter of the same circle
    (e) they all lie on the same line

56. A plane region that does not include its boundary is called
    (a) indefinite
    (b) non-Euclidean
    (c) open
    (d) closed
    (e) non-contiguous

57. Refer to Fig. Exam-7. The values a, b, and c are
    (a) the coordinates of a point on line L
    (b) the variables in the equation of line L
    (c) the solutions of line L
    (d) the direction numbers of line L
    (e) none of the above

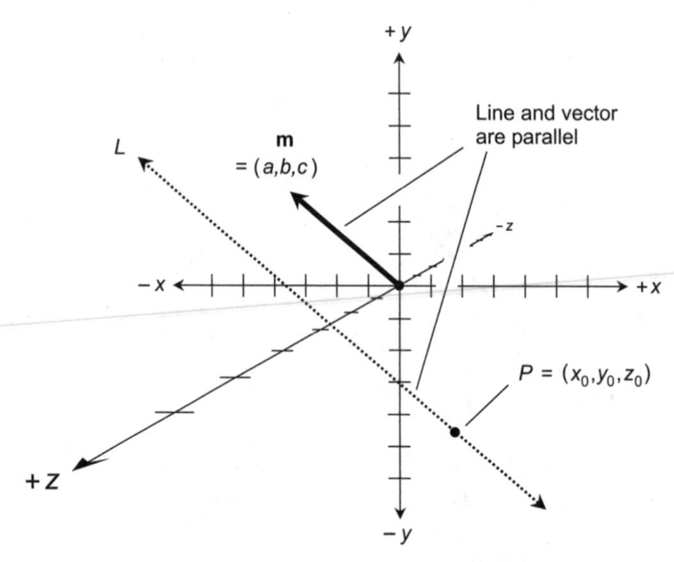

Fig. Exam-7. Illustration for Questions 57, 58, and 59 in the final exam.

58. In the situation shown by Fig. Exam-7, suppose the values of a, b, and c are all multiplied by -1. How will vector **m** and line L be related then?
    (a) They will be parallel to each other
    (b) They will be perpendicular to each other
    (c) They will be skewed with respect to each other
    (d) They will intersect each other
    (e) It is impossible to say without more information

59. In the scenario of Fig. Exam-7, what can we say about any point (x,y,z) on line L?
    (a) Its coordinates are $x = x_0 + a$, $y = y_0 + b$, and $z = z_0 + c$
    (b) Its coordinates are $x = x_0 - a$, $y = y_0 - b$, and $z = z_0 - c$
    (c) Its coordinates are $x = x_0 + at$, $y = y_0 + bt$, and $z = z_0 + ct$, where t is a real number
    (d) Its coordinates are $x = tx_0$, $y = ty_0$, and $z = tz_0$, where t is a real number
    (e) None of the above

60. Which of the following (a), (b), (c), or (d), if any, represents the same point on the Cartesian plane as the ordered pair $(-0.5,1.7)$?
    (a) $(-5,17)$
    (b) $(1.7,-0.5)$
    (c) $(-1/2,17/10)$
    (d) $(5,-17)$
    (e) None of the above

61. Let θ be the measure of an interior angle in a regular polygon. What is the range of possible values for θ?
    (a) $0 \text{ rad} \leq \theta \leq 2\pi \text{ rad}$
    (b) $0 \text{ rad} < \theta < 2\pi \text{ rad}$
    (c) $0 \text{ rad} \leq \theta \leq \pi/2 \text{ rad}$
    (d) $0 \text{ rad} < \theta < \pi/2 \text{ rad}$
    (e) None of the above

62. The distance between $(2,3,4,5)$ and $(6,7,8,9)$ in Cartesian 4D hyperspace is
    (a) equal to 2
    (b) equal to 4
    (c) equal to 8
    (d) equal to 16
    (e) equal to 32

63. Suppose a geometric object in the mathematician's polar coordinate plane is represented by the equation $\theta = \pi/4$. The object is
    (a) a circle
    (b) a hyperbola
    (c) a parabola
    (d) a straight line
    (e) a spiral

64. Imagine a stationary circle that has always existed, exists at this moment, and always will exist in the future. In 4D Euclidean time-space, the path of this circle, not including the points in its interior, is
(a) a hollow, infinitely long cylinder
(b) a solid, infinitely long cylinder
(c) a hollow sphere
(d) a solid sphere
(e) none of the above

65. Refer to Fig. Exam-8. What is the polar equation of the circle shown in this graph?
(a) $x^2 + y^2 + a^2$
(b) $x^2 - y^2 = a^2$
(c) $r = a$
(d) $r = \theta$
(e) $\theta = a$

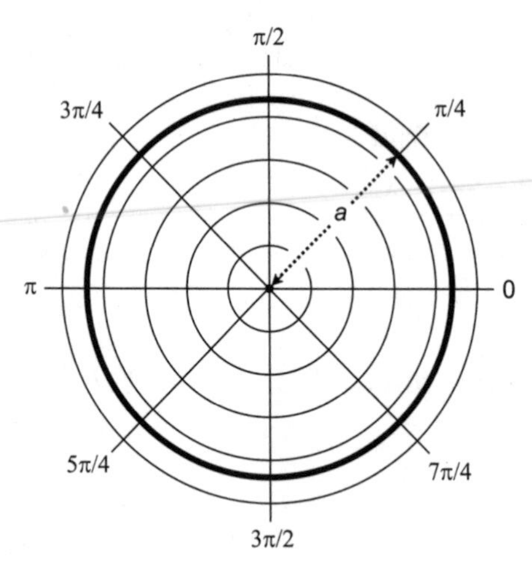

Fig. Exam-8. Illustration for Questions 65, 66, and 67 in the final exam.

66. Refer to Fig. Exam-8. What is the Cartesian-plane equation of the circle shown in this graph?
(a) $x^2 + y^2 = a^2$
(b) $x^2 - y^2 = a^2$
(c) $r = a$
(d) $r = \theta$
(e) $\theta = a$

67. Refer to Fig. Exam-8. Suppose the radius of the circle is doubled. What is the equation of the resulting circle in polar coordinates?
    (a) $x^2 + y^2 = 4a^2$
    (b) $x^2 - y^2 = 4a^2$
    (c) $r = 2a$
    (d) $r = 2\theta$
    (e) $\theta = 2a$

68. Suppose a line segment is bisected, and then the resulting line segments are bisected, and then the resulting line segments are bisected. How many times can this process, in theory, be repeated?
    (a) Until the bisection produces individual points
    (b) Forever
    (c) It depends on the length of the initial line segment
    (d) It depends on whether or not the end points are considered part of the initial line segment
    (e) It depends on whether or not the initial line segment is straight

69. Suppose there are two spheres S_1 and S_2, and the radius of S_2 is exactly four times the radius of S_1. Suppose A_1 is the surface area of S_1, and A_2 is the surface area of S_2. Which, if any, of the following equations (a), (b), (c), or (d) is true?
    (a) $A_2 = 4A_1$
    (b) $A_2 = 8A_1$
    (c) $A_2 = 16A_1$
    (d) $A_2 = 32A_1$
    (e) None of the above equations (a), (b), (c), or (d) is true

70. Suppose X is a plane, and P is a point. How many lines can exist that are normal to plane X and that pass through P?
    (a) None
    (b) One
    (c) Two
    (d) Infinitely many
    (e) It depends on whether or not P is in plane X

71. In a right circular cone, a line segment that connects the apex (top) and the center of the base
    (a) has a length equal to the radius of the base
    (b) has a length equal to half the radius of the base
    (c) has a length equal to twice the radius of the base
    (d) has a length equal to the circumference of the base
    (e) is perpendicular to the plane containing the base

72. Suppose you draw a line L and a point P near that line. Then you drop a perpendicular from point P to line L, and let Q be the point where the perpendicular intersects the line. Then you draw a point R on line L, different from point Q. The points P, Q, and R lie at the vertices of
    (a) a right triangle
    (b) an equilateral triangle
    (c) an isosceles triangle
    (d) a congruent triangle
    (e) none of the above

73. The distance between two parallel planes
    (a) is expressed along lines contained within both planes
    (b) is expressed along lines normal to both planes
    (c) is expressed along lines that intersect neither plane
    (d) varies with location
    (e) cannot be defined

74. Let L be a line parallel to a plane X. How many lines can exist in plane X that are skew to line L?
    (a) None
    (b) One
    (c) Two
    (d) Three
    (e) Infinitely many

75. A pyramid with a square base has
    (a) four faces in all
    (b) five faces in all
    (c) four or five faces in all
    (d) slant faces that are all rectangles
    (e) faces that are all congruent

76. If rotational sense is an important consideration in the expression of an angle θ, the counterclockwise sense usually indicates
    (a) $\theta = 0°$
    (b) $\theta < 0°$
    (c) $\theta > 0°$
    (d) $-\pi \text{ rad} < \theta < \pi \text{ rad}$
    (e) $-180° < \theta < 180°$

77. In the dimensionally reduced illustration Fig. Exam-9, the pitch of the cone (that is, the angle between the cone surface and the $+t$ axis) depicts

(a) the path of a single photon through time-space
(b) the hyperspace locations of the photons that came from the bulb at the instant it was switched on
(c) the hyperspace locations of all the photons that have come from the bulb since the instant it was switched on
(d) the speed of light
(e) the rate at which the observer travels through time

78. In the dimensionally reduced illustration Fig. Exam-9, imagine some plane X that is parallel to the xy-plane and that passes through the cone, so the set of points representing the intersection between plane X and the cone surface (not including the interior of the cone) is a circle. This circle represents
(a) the path of a single photon through time-space
(b) the hyperspace locations of the photons that came from the bulb at the instant it was switched on
(c) the hyperspace locations of all the photons that have come from the bulb since the instant it was switched on
(d) the speed of light
(e) the rate at which the observer travels through time

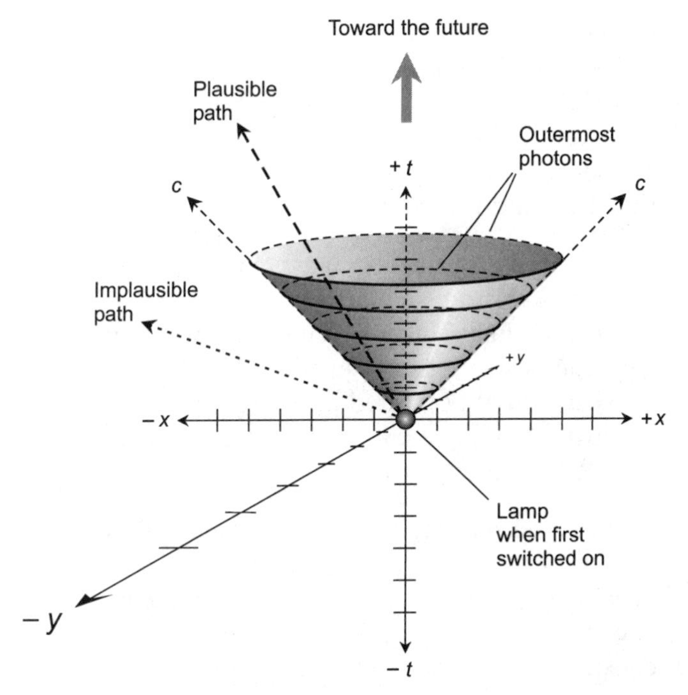

Fig. Exam-9. Illustration for Questions 77, 78, and 79 in the final exam.

79. In the dimensionally reduced illustration Fig. Exam-9, imagine some plane X that is parallel to the xy-plane and that passes through the cone, so the set of points representing the intersection between plane X and the cone (including the interior of the cone) is a disk. This disk represents
    (a) the path of a single photon through time-space
    (b) the hyperspace locations of the photons that came from the bulb at the instant it was switched on
    (c) the hyperspace locations of all the photons that have come from the bulb since the instant it was switched on
    (d) the speed of light
    (e) the rate at which the observer travels through time

80. Which of the following equations represents a parabola in Cartesian coordinates?
    (a) $y = 3x$
    (b) $x^2 + y^2 = 1$
    (c) $x^2 - y^2 = 1$
    (d) $y = 3x^2 + 2x - 5$
    (e) $x = -2y + 5$

81. A triangle cannot be both
    (a) isosceles and equilateral
    (b) isosceles and right
    (c) acute and obtuse
    (d) Euclidean and equilateral
    (e) Eucidean and isosceles

82. An uncalibrated drafting compass and a pencil, without a straight edge, can be used to
    (a) construct a line segment connecting two defined points
    (b) construct a line segment passing through a single defined point
    (c) construct an arc centered at a defined point
    (d) construct a triangle connecting three defined points
    (e) none of the above

83. An uncalibrated straight edge and a pencil, without a compass, can be used to
    (a) drop a perpendicular to a line from a defined point not on that line
    (b) construct an arc centered at a defined point
    (c) construct an arc passing through a defined point

(d) construct a parallel to a line, passing through a defined point not on that line
(e) none of the above

84. Imagine a triangle with interior angles measuring 30°, 60°, and 100°. What can be said about this triangle?
(a) It must be a right triangle
(b) It must be a non-Euclidean triangle
(c) It must be an isosceles triangle
(d) It must be a congruent triangle
(e) It must be an acute triangle

85. Consider an arc of a circle measuring 2 radians. Suppose the radius of the circle is 1 meter. What is the area of the circular sector defined by this arc?
(a) $^1/_2$ square meter
(b) 1 square meter
(c) $1/\pi$ square meter
(d) $2/\pi$ square meter
(e) π square meters

86. Suppose that a straight section of railroad crosses a straight stretch of highway. The acute angle between the tracks and the highway center line measures exactly 70°. What is the measure of the obtuse angle between the tracks and the highway center line?
(a) This question cannot be answered without more information
(b) 70°
(c) 90°
(d) 110°
(e) 290°

87. Suppose we are told two things about a quadrilateral: first, that it is a rhombus, and second, that one of its interior angles measures 70°. The measure of the angle adjacent to the 70° angle is
(a) 20°
(b) 70°
(c) 90°
(d) 150°
(e) none of the above

88. Suppose the coordinates of a point in the mathematician's polar plane are specified as $(\theta,r) = (-\pi/4,2)$. This is equivalent to the coordinates
(a) $(\pi/4,2)$
(b) $(3\pi/4,2)$

(c) $(5\pi/4,2)$
(d) $(7\pi/4,2)$
(e) none of the above

89. Suppose the cylindrical coordinates of a certain object in the sky are specified as (θ,r,h), where θ is its azimuth as expressed in the plane of the horizon, r is its horizontal distance from us (also called its distance downrange), and h is its altitude with respect to the plane of the horizon. Imagine that the object flies directly away from us, so r is doubled but θ remains constant. What happens to h?
(a) It does not change
(b) It doubles
(c) It becomes four times as great
(d) If becomes half as great
(e) It becomes one-quarter as great

90. Suppose the cylindrical coordinates of a certain object in the sky are specified as (θ,r,h), where θ is its azimuth as expressed in the plane of the horizon, r is its horizontal distance from us (also called its distance downrange), and h is its altitude with respect to the plane of the horizon. Imagine that the object flies straight up vertically into space, perpendicular to the plane containing the horizon, so h increases without limit. What happens to θ and r?
(a) Both θ and r remain unchanged
(b) θ approaches 90°, while r increases without limit
(c) θ remains unchanged, while r increases without limit
(d) θ increases without limit, while r remains unchanged
(e) It is impossible to answer this without more information

91. What is the slope m of the graph of the equation $y = 3x^2$?
(a) $m = 3$
(b) $m = -3$
(c) $m = 1/3$
(d) $m = -1/3$
(e) None of the above

92. Suppose we are told that a plane quadrilateral has diagonals that bisect each other. We can be certain that this quadrilateral is
(a) a square
(b) a rhombus
(c) a rectangle
(d) a parallelogram
(e) irregular

93. Suppose L is a line and P is a point not on L. Then there is one, but only one, line M through P, such that M is parallel to L. This statement is an axiom that holds true on
   (a) the surface of a flat plane
   (b) the surface of a sphere
   (c) any surface with positive curvature
   (d) any surface with negative curvature
   (e) any surface

94. Refer to Fig. Exam-10. Suppose two rays intersect at point P (drawing A). You set down the non-marking tip of a compass on P, and construct an arc from one ray to the other, creating intersection points R and Q (drawing B). Then, you place the non-marking tip of the compass on Q, increase its span somewhat from the setting used to generate arc QR, and construct a new arc. Next, without changing the span of the compass, you set its non-marking tip on R and construct an arc that intersects the arc centered at Q. Let S be the point at which the two arcs intersect (drawing C). Finally, you construct ray PS, as shown at D. Which of the following statements (a), (b), (c), or (d) is true?

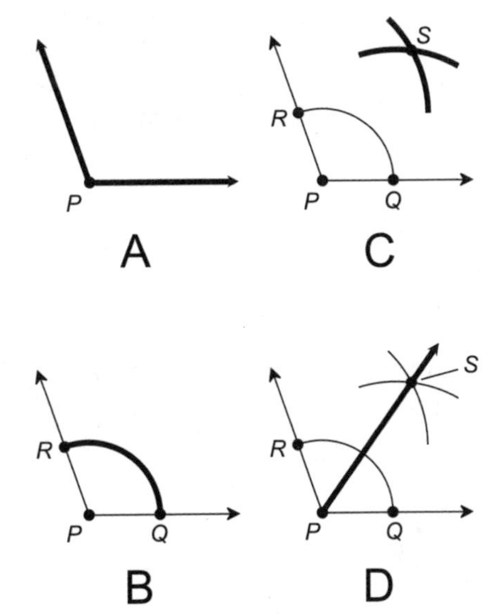

Fig. Exam-10. Illustration for Questions 94, 95, and 96 in the final exam.

   (a) $\angle RPS$ and $\angle SPQ$ have equal measure
   (b) $\angle PQS$ is a right angle
   (c) $\angle PRS$ is a right angle

(d) Line segment PS is twice as long as line segment RQ
(e) None of the above statements (a), (b), (c), or (d) is true

95. In the situation shown by Fig. Exam-10, and according to the description given in the previous question, which of the following statements is true?
(a) Points R, P, and Q lie at the vertices of an isosceles triangle
(b) Points R, P, and S lie at the vertices of a right triangle
(c) Points R, P, and Q lie at the vertices of an equilateral triangle
(d) Quadrilateral $RPQS$ is a trapezoid
(e) Quadrilateral $RPQS$ is a parallelogram

96. In the situation shown by Fig. Exam-10, and according to the description given in the previous question, consider the triangle whose vertices are points R, P, and S. Also consider the triangle whose vertices are points P, Q, and S. These two triangles
(a) are directly congruent
(b) are inversely congruent
(c) are both right triangles
(d) are both isosceles triangles
(e) do not resemble each other in any particular way

97. Refer to Fig. Exam-11. The perimeter, B, of quadrilateral $PQRS$ is given by which of the following formulas?
(a) $B = 2d + 2e$
(b) $B = ed$
(c) $B = 2f + 2g$
(d) $B = fg$
(e) None of the above

98. Refer to Fig. Exam-11. Suppose line segments PQ and SR are parallel to each other, and the line segment whose length is m bisects both line segments PS and QR. Based on this information, which of the following is true?
(a) $m = fd/2$
(b) $m = (f + d)/2$
(c) $m = eg/2$
(d) $m = (e + g)/2$
(e) None of the above

99. Refer to Fig. Exam-11. Suppose line segments PQ and SR are parallel to each other, and the line segment whose length is m bisects both line segments PS and QR. Based on this information, which of the following statements (a), (b), (c), or (d) is not necessarily true?
    (a) Quadrilateral $PQRS$ is a trapezoid
    (b) The line segment whose length is m is parallel to line segments PQ and SR
    (c) The distances e and g are equal
    (d) The distance h cannot be greater than the distance e or the distance g
    (e) All of the statements (a), (b), (c), and (d) are true

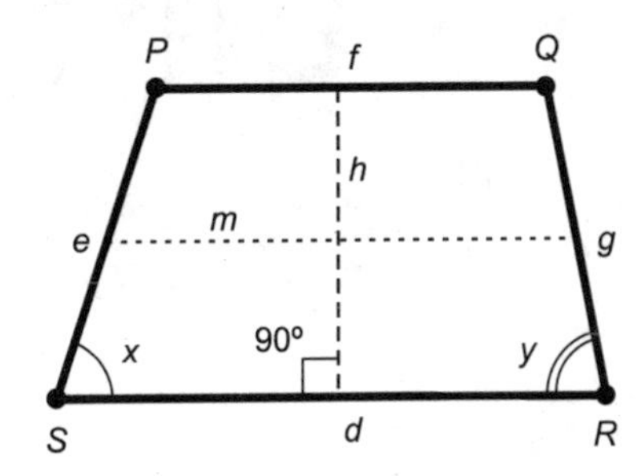

Fig. Exam-11. Illustration for Questions 97, 98, and 99 in the final exam.

100. Which of the following criteria can be used to establish the fact that two triangles are directly congruent?
    (a) All three corresponding sides must have equal lengths
    (b) All three corresponding angles must have equal measures
    (c) The Pythagorean theorem must hold for both triangles
    (d) Both triangles must be right triangles
    (e) Both triangles must have the same perimeter

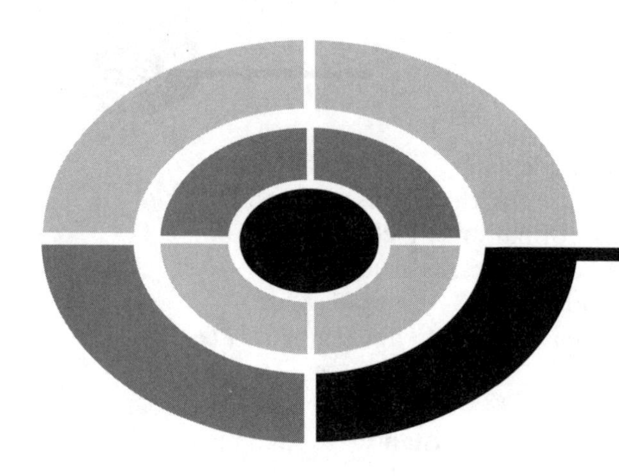

Answers to Quiz, Test, and Exam Questions

CHAPTER 1

1. c 2. a 3. d 4. a 5. d
6. b 7. d 8. d 9. c 10. c

CHAPTER 2

1. a 2. b 3. b 4. c 5. c
6. b 7. d 8. c 9. a 10. d

CHAPTER 3

1. c 2. d 3. c 4. c 5. a
6. d 7. b 8. d 9. a 10. b

CHAPTER 4

1. b 2. d 3. a 4. a 5. b
6. a 7. b 8. c 9. c 10. c

CHAPTER 5

1. c 2. b 3. d 4. d 5. c
6. a 7. c 8. a 9. d 10. d

CHAPTER 6

1. c 2. b 3. a 4. d 5. c
6. b 7. d 8. c 9. b 10. a

TEST: PART ONE

1. d 2. c 3. b 4. a 5. e
6. e 7. d 8. c 9. c 10. c
11. c 12. d 13. a 14. e 15. b
16. e 17. e 18. c 19. b 20. e
21. a 22. c 23. a 24. d 25. c
26. b 27. d 28. d 29. d 30. a
31. c 32. c 33. e 34. c 35. b
36. b 37. d 38. e 39. c 40. c
41. d 42. a 43. e 44. d 45. c
46. c 47. c 48. a 49. b 50. c

CHAPTER 7

1. c 2. a 3. c 4. d 5. b
6. a 7. a 8. d 9. c 10. d

CHAPTER 8

1. a 2. a 3. d 4. d 5. d
6. d 7. b 8. b 9. a 10. c

CHAPTER 9

1. c 2. a 3. a 4. b 5. b
6. c 7. c 8. d 9. d 10. a

CHAPTER 10

1. a 2. b 3. c 4. c 5. d
6. a 7. c 8. b 9. a 10. b

CHAPTER 11

1. b 2. d 3. a 4. b 5. a
6. c 7. d 8. b 9. a 10. c

TEST: PART TWO

1. e 2. b 3. c 4. e 5. d
6. c 7. b 8. a 9. d 10. c
11. c 12. b 13. d 14. c 15. a
16. a 17. a 18. e 19. d 20. e
21. c 22. c 23. a 24. a 25. b
26. b 27. d 28. e 29. d 30. e
31. e 32. d 33. a 34. b 35. e
36. e 37. d 38. e 39. a 40. e
41. c 42. e 43. d 44. b 45. c
46. b 47. d 48. c 49. e 50. c

FINAL EXAM

1. d 2. c 3. b 4. b 5. c
6. b 7. a 8. c 9. b 10. c
11. e 12. a 13. a 14. d 15. b
16. c 17. c 18. a 19. d 20. c
21. b 22. c 23. c 24. d 25. b
26. a 27. b 28. c 29. c 30. e
31. c 32. c 33. e 34. b 35. c
36. e 37. e 38. b 39. c 40. d

41. a 42. c 43. a 44. c 45. b
46. b 47. c 48. d 49. e 50. a
51. c 52. e 53. e 54. c 55. e
56. c 57. d 58. a 59. c 60. c
61. e 62. c 63. d 64. a 65. c
66. a 67. c 68. b 69. c 70. b
71. e 72. a 73. b 74. e 75. b
76. c 77. d 78. b 79. c 80. d
81. c 82. c 83. e 84. b 85. b
86. d 87. e 88. d 89. b 90. a
91. e 92. d 93. a 94. a 95. a
96. b 97. e 98. b 99. c 100. a

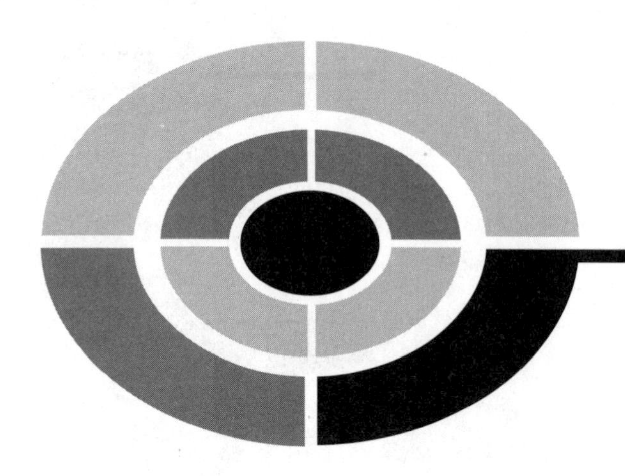

Suggested Additional References

Books

Arnone, Wendy, *Geometry for Dummies*. New York, Hungry Minds, Inc., 2001.

Gibilisco, Stan, *Trigonometry Demystified*. New York, McGraw-Hill, 2003.

Huettenmueller, Rhonda, *Algebra Demystified*. New York, McGraw-Hill, 2003.

Leff, Lawrence S., *Geometry the Easy Way*. Hauppauge, NY, Barron's Educational Series, Inc., 1990.

Long, Lynnette, *Painless Geometry*. Hauppauge, NY, Barron's Educational Series, Inc., 2001.

Prindle, Anthony and Katie, *Math the Easy Way*. Hauppauge, NY, Barron's Educational Series, Inc., 1996.

Web Sites

Encyclopedia Britannica Online, www.britannica.com.

Eric Weisstein's World of Mathematics, www.mathworld.wolfram.com.

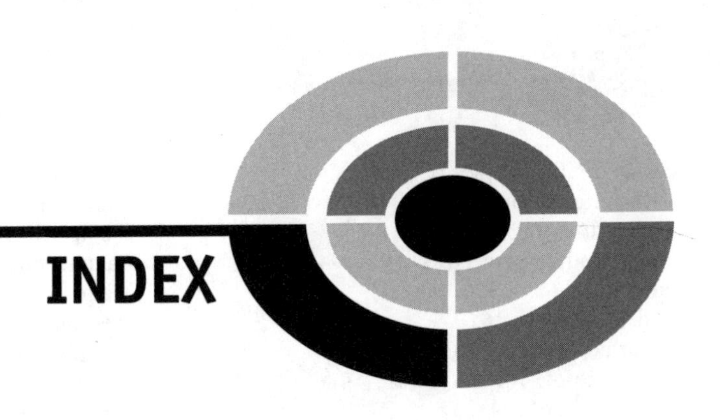

INDEX

ABOUT THE AUTHOR

Stan Gibilisco is the author of many best-selling McGraw-Hill titles, including the *TAB Encyclopedia of Electronics for Technicians and Hobbyists, Teach Yourself Electricity and Electronics*, and *The Illustrated Dictionary of Electronics*. *Booklist* named his *McGraw-Hill Encyclopedia of Personal Computing* one of the "Best References of 1996."